ANIMAL RIGHTS Vs. NATURE

Walter E. Howard

Available from:

Bookstores or a check or money order payable to:

Mr. J. Casey Howard
Rt. 1, Box 59
Baker City, Oregon 97814
503 523-3383, 800 627-2530

$ 10.00 plus $ 2.00 handling charge.

Published by:

Dr. Walter E. Howard
24 College Park
Davis, California 95616

First Edition, July 1990

Second printing 1991

©1990 by Walter E. Howard

All rights reserved. No part of this book may be reproduced in whole or in part in any form or by any means, electronic or mechanical, including photocopying, recording, or by any information storage and retrieval system, without permission in writing from the author: Department of Wildlife and Fisheries Biology, University of California, Davis, California, 95616.

Library of Congress Catalog Card Number: 2 939 274

ISBN 0-9627641-0-8

Printed in the United States of America.

TO OUR ANIMAL FRIENDS

MAY ANIMAL WELFARISM AND TRUE COMPASSION FOR ANIMALS PREVAIL OVER THE EXTREME ANIMAL RIGHTS MOVEMENT OF "NO DOMESTIC LIVESTOCK, POULTRY, PETS, OR LABORATORY ANIMALS," WHICH INEVITABLY WILL ALSO CAUSE THE LOSS OF MANY GAME ANIMALS AND FURBEARERS.

Credits

I am grateful to many people for their assistance, especially that received from Dr. Werner T. Flueck and Dorothy Beadle. The California ground squirrel on the cover was drawn by Linda D. Heath of the California Department of Food and Agriculture. Rex E. Marsh provided most of the original drawings which were done by students. They were redrawn for this book by Claudia Graham. Craig Hillis redrew the bald eagle, which was photographed by Joseph Von Wormer.

CONTENTS

	Preface	
1.	Introduction	1
2.	Nature	21
3.	Animal Rights	55
4.	Vested Interests	79
5.	Conservation Biology	93
6.	Wildlife Management Animal Control	103
7.	Hunting and Trapping	131
8.	Fur	157
9.	Domesticated Animals	163
10.	Animals In Research	171
11.	Human Population	179
12.	Conclusion	193
13.	References	203
14.	Appendix	213

The national bird of United States of America, the bald eagle (<u>Haliaeetus</u> <u>leucocephalus</u>), has made a remarkable recovery primarily as a result of protection as an endangered species, but also due to banning DDT. (Original photo Courtesy of Joseph Van Wormer.)

PREFACE

I am highly concerned about animal welfare and have devoted most of my 43 years of research and teaching as a professor at the University of California, Davis, trying to find better ways for people and animals to cohabit (live close to each other). I don't question the sincerity of many followers of the animal rights movement but, for reasons developed in this book, I wish their energies could be redirected so as to be more productive in helping animals.

This is a forthright and objective analysis of the common misconceptions many people have about the balance of nature in environments altered by civilization. You will probably be shocked to find that many of your cherished, almost romantic, views about nature and the animal rights movement are false, being based on emotional propaganda rather than supported by biological facts. You may not always agree with the author, but you will certainly be stimulated to rethink your own perspectives. The book is excellent for motivating students to think. The book is not just for those interested in humaneness and animal rights, but covers subjects in the fields of ecology, wildlife conservation, biomedical research, animal control, agriculture, hunting and trapping, political science, and the social sciences.

The hardcore activists of the animal rights movement want to dress you their way, make you a vegetarian, and provide a world devoid of pets, livestock, and many species of game and furbearers. Concerning the wild animals, this book will show how it often is the animal rights movement that is inflicting the greatest cruelty on animals and not sportsmen.

It seems I am a likely candidate to analyze these issues because of my wide background of research, teaching, and travel/lecturing on how animals and people can best coexist. The courses I have taught include wildlife ecology, principles of animal control, population problems/issues in human ecology, and animal welfare. In addition, 50 of my foreign travels have been on lecturing/consulting assignments. Life of any organism is complicated. But isn't it wonderful today that so many people want to help animals and think we should stop inhumane exploitation of animals. This book takes a critical look at the difference between the animal rights movement and animal welfare, and whether or not the laws of nature are being accommodated. I want the public to be properly informed about animal rights by better understanding the legitimate concerns about the role of nature in animal welfare so they can make constructive decisions instead of being influenced so strongly by emotional propaganda.

If you are an ardent animal rightist supporter, you should not read this book, for it may weaken your convictions. Regardless of your philosophy about human exploitation of wildlife, this book will be a strong stimulus and force you to develop more eloquent arguments as to why your views are ethically sound.

Since writing <u>Nature and Animal Welfare: Both are Misunderstood</u>, I have given many lectures in the U.S. and overseas. I learned a great deal from my audience when I gave the Universities Federation for Animal Welfare's Hume Memorial Lecture (Nature's Role in Animal Welfare) at The Royal Society of Medicine in London on 29 November 1989, again at The Royal Society of Edinburgh on 1 December, and then spoke on a related subject at the University of Reading. All of my audiences have been a great help in the preparation of this book. My mission is to make people think more deeply about the issues I raise, admittedly with a broad brush, in search of a more compassionate universe.

Many people have been frustrated by what I say because they are perplexed as to how to respond. Certainly, they do not go to sleep during my lectures. I have also been privileged to read the book reports of students in classes that have been assigned my earlier book, and without exception it has stimulated them all to think more deeply about this subject, whether or not they agreed with me on some parts.

People need a refuge from their everyday business woes, personal pressures, and what might aptly be called the psychic guerilla warfare of daily life. But how do we preserve breathtaking outstanding natural wonders like upper Yosemite Falls in Yosemite National Park, while at the same time permitting large numbers of people to enjoy the experience? (W.E. Howard)

1. INTRODUCTION

The goal of this book is to encourage more humane treatment in the use of animals, with special emphasis on wildlife in modified environments. Animals do not need sympathy, they need compassion. They need to be properly understood and managed according to the rules of nature. Nature needs to be brought back into the deliberations of animal rights. If people did not enjoy and use animals as pets, beasts of burden, for sport, for food, and for products, a large percentage of the vertebrate animals we see today would not have been born and many species would be extinct. What is morally wrong with exploiting animals for human purposes, if done humanely? All animals exploit other animals, and humans are the only ones that try to do it without inflicting unnecessary suffering.

To me, it is wonderful that so many game, research and domestic animals get to be born, if we can assure that they will live a healthy quality life. It is unfortunate that their lives may be terminated prematurely, but such utilization means that even a greater number of animals will be born. Such a process is acceptable as long as their death comes as humanely as possible, i.e., without the pain and suffering nature usually inflicts when terminating a life.

Animal Rights Vs. Nature

We must recognize that it often is necessary to inflict some pain and suffering in the management and use of animals, and I hope to show why such suffering is not only acceptable but often is desirable if the suffering is kept to a minimum and not done unnecessarily. Today many wild animal populations in disturbed environments lack an adequate natural or human-caused mortality factor and as a consequence suffer horribly from species' self-limiting factors such as starvation, disease, and cannibalism.

INTENTIONAL MISTREATMENT OF ANIMALS WITHOUT MORAL OR ETHICAL CONSIDERATIONS SHOULD NOT BE TOLERATED.

My objectives for writing this book are not to ridicule and discredit the estimated 6% of the Americans who belong to animal rights groups. Rather, my goal is to air what I think are their misjudgments. Animal rightists have done a lot of good in creating awareness to the need for humane treatment of animals, but they have caused much suffering of animals and their emotionalism has ignored the laws of nature. Due to a few terrorists and extreme environmentalists, the basic image of the animal rights movement has been severely tarnished. Likewise, some of the leaders have not helped where they make extreme statements approaching nincompoopery, as quoted in The Washingtonian: "Human beings have no right to the knowledge gained from experimentation on animals -- even if done painlessly." "I don't believe human

Introduction

beings have the right to life. That's a supremacist perversion. A rat is a pig is a dog is a boy. Meat-eating is 'primitive, barbaric, and arrogant,' and pet ownership is fascism." "The time will come when we look upon the murder of animals as we now look on the murder of men."

The dedication of most animal rightists is sincere. I have attended their seminars and conferences. It is clear that many are religious vegetarians who thrive on what they think is wrong, but are unobjective idealists who stress emotionalism and are not interested in objective debates. But what is needed are open, constructive forums so that the best of all philosophies can be brought to the mainstream. They do not discuss the laws of nature. I may be fighting a losing battle, but I hope to expedite the process of bringing animal welfare, not animal rights, to the forefront. Where I am wrong, I want to be set straight for I too am highly concerned about the welfare of animals and the environment. I believe that we need stronger social ethics in support of **ANIMAL WELFARE**, not animal rights.

We cannot and should not discard the results of thousands of years of cultural developments, although there is a need to make adjustments to the current thinking and behavior. Professional experts nowadays quite often have to spend more time correcting or minimizing damage caused by well-meaning, enthusiastic but incompetent idealists, than working in their area of expertise. Even more dangerous are such wellmeant movements when they combine their main goals with nonscientific goals such as rights for minorities, liberation, religious or political idealogies. In all such cases the majority of followers is driven by good intentions and enthusiasm, but basically such movements reflect the absence of a real concept of the whole.

NATURE CAN BE THE ENEMY.

Those of us who exploit animals either as pets or for food, labor, or their products, or use them in research have been slow to adjust to needed changes in social attitudes for animal welfare; hence, the explosion of the animal rights movement. The improvements needed in animal welfare opened the door for the current extreme animal rights activism. Since society always has had a small core of anti-establishment, intellectual, loner, reactionaries who gravitate to opportunities for activism and show defiant resistance to excessive bureaucratic authority, the stage was set for new leadership to exploit.

The unfortunate trend through the 1980s was the blind belief that the animal rights movement would soon pass if ignored. The conservatives remained mum instead of counteracting the activism. Now the mistake has been realized and most recognize that society's social attitude toward animal welfare is evolving rapidly and continuation of these philosophical changes seems inevitable. However, the public has also begun to believe too much of the activists' emotional propaganda, even though laws of nature (to be discussed later)

Introduction

have been ignored in the propaganda. Because the media profits form sensationalism, activists have had a heyday, often by breaking the law, claiming their activities were civil disobedience done for just causes. However, to be honest, we all now have a greater concern for the welfare of animals because of the animal rights movement.

ANIMAL RIGHTS MOVEMENT HAS STIMULATED SOME HEALTHY SOUL SEARCHING.

It is not easy to transcend the emotional barriers encountered at every turn when trying to be objective about nature's role in the animal rights issues when you love wildlife and have a religious respect and awe for the environment. I also care about the air I breathe, water I drink, and the fate of the soil that provides my sustenance. During my 43 years of research and teaching I have searched for more humane and acceptable ways animals and humans might cohabit (live together) in **MODIFIED** environments. This has made it necessary for me to carefully examine ethical questions about nature's role in animal rights, whether it be how pets and farm animals are treated, use of laboratory animals in research, or the hunting and trapping of animals for sport or subsistence.

As an ardent nature lover and naturalist who has had wide experiences, including 50 lecturing or consulting assignments in foreign countries, I believe I have a good understand-

Animal Rights Vs. Nature

ing and appreciation of nature. It appears that most animal rights groups ignore many of the laws of nature, hence the public steadily distances itself from the true realities of nature. Upon reading this book, it will become obvious that I am not an animal rightist; however, I hope it will also be evident that I am a strong supporter of animal welfare. In addition to sensible animal welfare, I am interested in other movements such as anti-apartheid, the women's movement, and the need to protect the global fauna and environment, but will not discuss them here. I received a Ph.D. in Vertebrate Ecology in 1947. I feel I am in a unique position, being an emeritus professor with a very broad natural history, agricultural, research, and teaching background. Relevant courses I have taught include wildlife ecology, principles of animal control, and animal welfare. In this book I'll strive to keep the theoretical nature of my animal welfare viewpoints in context.

As a nature lover and naturalist, I join many other concerned environmentalists, conservationists, and wildlife managers in being pleased with how many nonanimal research models are being developed, the increase in research to design more humane animal traps, and the increased humanism with pets and livestock. And I also join these folks in being displeased about the unnecessary inhumane existence many animals and wildlife populations sometimes are forced to experience as a consequence, perhaps inadvertently, of actions taken by those who may think they are helping wild animals by not letting people aid nature by managing the fauna in altered environments. The professionals in wildlife management and wildlife conservation are under siege by some

Introduction

outspoken animal rights and environmental extremist groups. Who is right, and when is each side overstepping its boundaries?

This is not a review of the subject of animal rights, for it would be unfair to cite competent authors out of context as if they endorsed my views, which clearly do not parrot the general consensus regarding the animal rights issues. This book will examine some of the emotional issues that have spawned the creation of numerous extremist environmental and animal rights organizations, and how such groups have been able to prosper. Most environmental and conservation organizations, of course, are valuable assets. Without genuine concerns about animal welfare, the various new laws concerning the care and research use of animals in laboratories and outdoors would not have come about so quickly.

Balance of nature issues have become so emotional that even many scientists find it difficult to discuss these matters objectively, and academia often emotionally preaches rather than teaches the controversial natural principles involved in the animal rights movement. They often don't stimulate others to think objectively; instead, try to make them accept their own views. The animal rights movement has not been adequately challenged partly because the informed biologists are under organizational restrictions against speaking out. These facts, along with the extreme adversary approach taken by many environmental organizations with their strong <u>vested interests</u> in "pot stirring" with "cheap shots" do not always benefit animal welfare or help the environment.

Animal Rights Vs. Nature

To help counteract the extreme animal rights groups, a new "death ethic" is urgently needed, because nature demands a high rate of premature deaths if healthy reproducing populations of wildlife are to prevail. In modified environments it is often a choice of either letting a survival-of-the-fittest new balance develop or become a predator to help "desirable" species, such as many endangered species, that have become less fit in the new environment.

A steadily growing percentage of the public no longer understands the basic biological principles responsible for the so-called balance of nature.

QUALITY OF LIFE REQUIRES AN EFFECTIVE MORTALITY FACTOR.

Animal rights are too complex and words too inadequate for anyone to elucidate nature's role in the survival and well-being of animals. The viewpoints each of us has about animals and their welfare are very personal and diverse, based for the most part on our heritage, religion, previous experiences, and education. But nature's role in animal welfare has been ignored by much of today's emotionalism and activist rhetoric concerning animal rights. A follow-the-leader mentality has evolved by people who are opposed to any treatment or management of animals the media tells them is bad without any attempt to analyze such treatment, hence smothering clear-thought processes. Consequently, this distortion has seriously

Introduction

polarized the issue of animal rights and produced a large segment of the concerned yet uninformed public that is ignorant of the laws of nature. Yet these are laws that we cannot change; hence, paradoxically, we must work with them if we are to create harmony in altered environments.

Unfortunately, all activists have to continue being more and more dramatic or else be forgotten. Some of my favorite environmental organizations have at times lost credibility by using this fund-raising, anti-establishment adversary approach. The tyranny of an uninformed and misled public is difficult to surmount, but with a better understanding of nature, the public may be able to look more closely at some of the solicitations for funds by these organizations. It is not always easy to discriminate between legitimate appeals and those by people who have to make their living from public donations. Such solicitations are often cleverly couched so as to ease the donor's conscience and most of us don't have time to study each issue. Because none of us can support them all, we must learn how to recognize emotional propaganda from sound biological and ecological facts. The more sensational the solicitations, the less likely is it that very much of your contribution will go to help animals.

ALL WILD AND DOMESTIC ANIMALS HAVE LEGITIMACY AND VALUE AND DESERVE HUMANE TREATMENT.

Animal Rights Vs. Nature

We often get hung up philosophically with personal ethics and values rather than trying to analyze the issue from a biological perspective. To say that higher animals are not human does not resolve the question, nor is humaneness the main issue. The only right nature seems to give animals is the right to try to survive. However, I think all of nature's animals have legitimacy, value, and sensory capabilities. This includes the dominant omnivore, people, who are part of nature. The psychological well-being of confined higher animals is also important, although it may be difficult to arrive at conclusive evidence of well-being.

I hope this book will challenge some of your present beliefs in order to have you think more deeply and objectively about the welfare of all animals and environmental issues, including some sound points raised by animal rightists. The book examines some of the widespread common misconceptions about the balance of nature, especially the role of nature in man-modified environments, but also considers the use of animals in research and the role of domesticated animals.

The goal of many people is the establishment of nature as it was before Europeans arrived EXCEPT for things that affect them. Humans do not eliminate natural balances; rather, they merely initiate new balances. Whether the new balances are better or worse depends upon one's viewpoint. What I am offering is an opposing value system of which my main tenet is that it is better for a population of animals living in man-disturbed environments to have the surplus individuals selectively killed by humans rather than to let most of the

Introduction

population become highly distressed or even die due to disease, starvation or other factors resulting from habitat alterations.

Because I think I may understand nature better than many, I cannot remain untouched by the sensationalism and polarization that some segments of the animal rights movement have recently cast upon respected animal welfare programs. Many people today are urban folks who have little experience of nature in the raw. They have not lived on the land, nor have they experienced the joys, triumphs, and misfortunes of producing crops, raising livestock, or otherwise having to be responsible for managing and harvesting food to feed the family. They usually don't know the odors of a dairy or piggery. Also, many have been deeply influenced by anthropomorphism and the "Bambi" syndrome that ignores nature's life-and-death roles, so how can they understand that to improve animal welfare in modified environments it is usually unwise to leave the outcome of any new balancing process to the whims of nature?

I don't think the public should be fooled into thinking that wild animals will live to a ripe old age if only protected from humans, and that wildlife have the "right" to live a long life. The reason wild animals seldom reach senility or an old age is because they don't have pesticides, our technology to provide them with safe water to drink, or safe, abundant food to eat, and ability to provide a safe indoor and outdoor environment in which to live. Also, few people realize that humans show more compassion toward other members of their

Animal Rights Vs. Nature

species and to other species than do wild animals.

No plant or animal, or any organism for that matter, "wants" to die. The emotional awareness and consciousness of higher animals are apparent, but remember the balance of nature demands that the bulk of all lower and higher organisms born must die prematurely. It is nature's food web. People are currently the only exception, and look what the burgeoning human population has done to much of the earth's environment, even causing the extermination of many animal species.

Public attitudes are probably more influential than laws and policies, and they are deep-rooted in issues of values and goals. I am assuming that if I can stimulate readers to think more objectively, there will be a chance of getting more people to base their attitudes about nature on biological principles and common sense. By creating better awareness of the true role of the balance of nature and the development of better alternatives to managing all types of problems with wildlife, perhaps society can then set more environmentally desirable goals concerning wildlife. Better answers are needed. Previous generations have not done the best job of protecting our wildlife heritage and, unless changes are made, the future doesn't look too bright.

We tend to overlook that people are part of nature, that they must be accommodated, and that the most basic law of nature is that each organism survives by consuming other forms of life. Scavengers and decomposers rely upon what is

Introduction

left of organisms slaughtered by others or which have died from disease, starvation, or other vicissitudes of life. But it is intolerable for people, the most intellectual animal, not to demonstrate compassion for the welfare of other animal species.

PEOPLE RESPOND EMOTIONALLY RATHER THAN COMPASSIONATELY.

It is not easy to think globally when wrestling with local issues. Instead of extreme reverence for life of individual wild animals, society needs to develop a "death ethic"; that is, a realization that it is a law of nature that large numbers of animals must and do die prematurely to enable a healthy population of each species to exist.

I take exception with those expressing a higher imperative view against any willful killing of animals. I support killing of animals as long as the doer is abiding by his or her ethics and not indiscriminately causing wanton brutality or unnecessary suffering; hence, the importance of education in developing responsible and healthy ethics. My argument, as I stress elsewhere, is that people can and often should play a more effective and humane mortality role than would occur if an altered environment is left to nature's way.

Animal Rights Vs. Nature

DESTROYING NATURAL MORTALITY FACTORS CAN BE A DEATH TRAP OF POPULATIONS.

Is it natural for people to enjoy animals as pets, beasts of burden, for sport, and for food and other products? If we didn't, most of these animals would not have been born. Horses would, in fact, probably be extinct if they had not been domesticated. Many people do not object to killing animals for food, even though we do not need meat to survive, yet object to killing animals for fur coats. Is there a difference? OK, so you are a vegetarian! That is not an escape, even though many animal rightists think the human race should become vegetarian. Vegetarians cannot live on plants growing naturally and untended. The grains, vegetables, fibers, and fruits they seek require local eradication of many wild animals, especially mammals, so these plants can be economically grown, whereas much meat can be produced by livestock and game utilizing various plants and plant materials that we cannot eat. But it is true that if we did not eat meat, theoretically we could feed more people in the world. More importantly, there has evolved in nature the existence of the higher forms of animals as meat eaters, and they are essential to preserve certain balances.

According to Garn and Leonard, hominoids (5-8 million years ago) were omnivorous, but mainly scavengers. The first <u>Homo erectus</u> (early Pleistocene) was a hunter. Climatic and faunal associations indicate that they were not browsers.

Introduction

<u>Homo</u> <u>sapiens</u> appears in late Pleistocene and was a more sophisticated hunter. Vegetation contributed increasingly to the diet, and resulted in the growing of grain. This was a prerequisite for population expansion demanding the cooperative behavior of entire groups. With the advent of villages, the size of people became reduced, and there are also indications of iron, vitamin C, and protein-energy deficiencies. "It is a part of our intellectual tradition to romanticize the past, both so-called Natural Man and an imaginary time in the past when people were presumably healthy and happy subsisting on unprocessed food."

If people were "natural" vegetarians they would have a large cecum instead of a rudimentary appendix or additional stomachs like ruminants. In the New England Journal of Medicine, Dr. John Lindenbaum, a Columbia University professor of medicine is quoted as saying that their study showed that vegetarians who did not take vitamin B_{12} supplements can suffer depression, memory loss, disorientation, sluggish mental processes, irritability, marked personality changes, agitation and abnormal behavior patterns. Therefore, strict vegetarians always develop vitamin B_{12} deficiency very slowly over a period of several years. The only natural sources of vitamin B_{12} in our diets come from meat, fish, eggs, and to a lesser extent dairy products. Animals cannot make B_{12}, and herbivores absorb B_{12} after digesting microorganisms containing the vitamin, according to Dr. Neal D. Bernard, President of Physicians Committee for Responsible Medicine (The Animals Agenda, March 1990). If you practice a vegetarian diet, you should make sure you eat foods enriched with B_{12}, i.e., a compound called cobalamin, or vitamin supplements such as spirulina. Furthermore, the amino acid composition in vegetation is not equivalent to animal protein. However, amino acid deficiencies can develop in humans through an inappropriately balanced vegetarian diet. In fact, it requires sophisticated plant protein analysis, and specifically designed vegetarian diets to ensure a proper balance of amino acids. It is doubtful that a strict vegetarian would be able to maintain a healthy diet in terms of amino acid balance without the advanced knowledge of human nutrition and the technology to continuously analyze plant proteins. Designed vegetarian diets consist of a specific mixture of plants to mutually supplement each other's amino acid composition, but purified amino acid supplements are also being used.

Arguments by vegetarians that when we eat beef we get only about 1/10 of the energy and protein we would get if we consumed their feed directly are not sound. Most of the food input of cattle, for example, consists of forage and byproducts that are not edible by humans. According to Bywater and Baldwin, at an American Association for the Advancement of Science Symposium, the amount of edible portions of a beef carcass returns about 5.1 percent of the energy and 5.3 percent of the protein consumed, and feed conversion of an animal from birth to slaughter is approximately 9.51. However, when the human palatability of the feed inputs is considered, they found more human edible protein was usually returned in the beef than was consumed by the livestock. In case of milk production, returns of both protein and energy may be greater than 100 percent of the edible inputs. As might be expected,

Introduction

returns of edible inputs from ruminants, like cattle, are higher than for nonruminant swine and poultry. Livestock of all kinds are primarily producers of high-quality protein, not energy.

Isn't it wonderful that most people in developed countries no longer have to struggle against many of the vagaries of nature which stymied early settlers' survival, and that we can find opportunities to enjoy the mental and spiritual rewards nature can provide us? But we must guard against letting our joyous behavior and treatment of animals be guided too selfishly. It is only natural to want to treat animals in a manner that will make us feel warm and glowing inside, even though too often we will not be showing true compassion for the animal or population in question. To try to protect all animals from a premature death or suffering may actually work against the very well-being and humane principles one is striving for.

SUFFERING IS A NATURAL PHENOMENON WITH CONSIDERABLE SURVIVAL BENEFIT.

Animals in the wild do not have access to morphine. Only some people have the benefits of pesticides, clean water, public health, modern medicine, and surplus nutritious food. Anthropomorphically, how important is an animal's life? Should we let as many as possible be born so they can experience life? Should more animals be bred for laboratory

Animal Rights Vs. Nature

research as long as their end comes humanely with minimal suffering? Likewise, why not greatly increase the number of fur farms? Such furbearers will almost all have a healthier life and die more humanely than would happen naturally. Their quality of life must be fairly good or the animals will not have good fur or reproduce successfully. Because they don't "know" what it is like to live free, there is a good chance they don't "mind" being reared in pens or on small islands.

This wooden telephone pole illustrates the ferocity of nature. On 24 April 1984 near Turpan, Xinjian Province, China experienced a severe Gobi Desert dust storm where wind-blown small stones etched away much of the pole and part of the concrete post support. Without adequate shelter, no wildlife could survive such a cruel storm. (W.E. Howard)

2. NATURE

What do we mean by the term "balance of nature"? The harmony or "balance" found in undisturbed nature is a dynamic biological equilibrium resulting from the sum of interactions among all resident organisms in their attempt to survive and reproduce. They sustain themselves by reproducing and eating each other. It is that web of relationships among the population densities of the diverse species of organisms that makes up an ecological community. The balance is ecosystem homeostasis, the intrinsic regulations, adjustments, or feedback mechanisms, which also creates new equilibriums (balances) in disturbed environments, leading to a climax or a more or less steady state between inputs and outflows of nutrients. From a practical point of view, the balance of nature should be called "balancing" of nature, for it is the dynamic struggle for existence, i.e., the survival of the fittest, in the cruel world of nature where it is the natural right of the largest and strongest (or smallest but quickest) to feed upon or displace the smallest or weakest (or largest but slowest), even if of the same species.

Any substantial human disturbance of an environment causes a shift in relative abundance of coexisting species and can result in a reduction of species richness. It is inevitable that our activities will alter the relationships of organisms to the environment. No species can live in all environments and once its habitat is destroyed the species will vanish from that area. However, habitats or environments cannot be eliminated, only altered, and nothing succeeds like biological succession. As J. A. MacMahom pointed out, new and better man-made ecologically and economically sound ecosystems are feasible. The so-called "harmony" or "balance" in nature is really no more than the current degree of biological equilibrium established by the sum of interactions among all the organisms present as each attempts to survive and reproduce.

What is natural? Is it a process, i.e., the consequence of nonhuman caused events? Unfortunately, it is an axiom that people, like all species, must exploit the environment, hence change the balancing components in order to survive. To sustain themselves in a balanced community, the animal populations must have high mortality rates to prevent excessive populations that might permanently alter their habitat.

ALL SPECIES ARE PROGRAMMED TO OVERPRODUCE; THE SURPLUS MUST DIE PREMATURELY.

In nature, all species, including people, are inherently programmed to overproduce. The successful species have displaced competitors that were not as well adapted. Species that survive when the habitats are altered by people may then have an impact on the changed environment that is either advantageous, deleterious, or neutral to the welfare of the other species. All attempts to help restabilize a community,

Nature

including no additional human action, no matter how valuable and desirable, may still have tradeoffs that we consider undesirable.

Both birth and death rates of most species of wildlife are high. For example, if any rodent had the same survival rates as humans, the species would soon cover the earth.

NATURE DEMANDS A HIGH PREMATURE DEATH RATE IF POPULATIONS ARE TO BE MAINTAINED IN A HEALTHY MANNER.

The prevention of natural deaths leads to overpopulation, disease, and starvation in wildlife species, just as it does in human populations or underharvested domestic animals. Natural mortality factors have evolved to restrict the distribution and density of animals in natural environments to match the available food supply, good harborage for breeding, and cover from predators. Carrying capacity is the number of animals the habitat can support and perpetuate through the year without damaging the future welfare of the animals or their habitat. When wildlife numbers are not controlled by mortality factors and exceed the carrying capacity, excess animals will eventually die but first they may damage the environment. Many national parks are experiencing the consequences of imbalance: elephants in Africa eliminated trees they needed for food, hippos caused serious erosion along riverbanks before poaching became so serious, and red

Animal Rights Vs. Nature

deer in Switzerland cause serious erosion because there are no large predators.

It is an axiom that man must exploit the environment and deliberately unbalance it to survive. The "man must not meddle with nature" philosophy needs to be dispelled because, 1) it is equivalent to denying humans to exist, and 2) to **"LEAVE IT TO NATURE" AFTER MAN HAS ALTERED THE ENVIRONMENT IS SELDOM A WISE SOLUTION, ECOLOGICALLY OR HUMANELY.** Few people realize that agricultural crops could not survive if all native mammals were treated like endangered species. In fact, most home landscaping and city parks would also be destroyed, at least aesthetically, if native mammals were allowed free range. Since these plants are exotics, they have not co-evolved to withstand the feeding pressures of the native mammals.

In modified environments, if one is willing to accept the consequences of the survival-of-the-fittest process and the loss of those species that now are unfit, then it is okay to let nature take its course. A common sequel is a reduction of productivity, and in many instances this would also cause the demise of sensitive, threatened, or endangered species and their habitats. In altered environments the choice is ours to either accept whatever new balance develops "naturally" or to humanize nature and help the species we consider desirable.

Pain and suffering are an integral part of nature and are essential to evolution and the survival-of-the-fittest process.

Nature

However, even though we are part of nature, this still does not give us a license to inflict "unnecessary" suffering; i.e., we should make the lethal management tools as humane as possible, yet still assist nature in modified environments by harvesting the surplus that otherwise would damage the environment and the species' own welfare.

Sometimes helping animals is not simple. When a caring but untrained person tries to give a surplus wild mammal that has been driven from its normal habitat another chance for life by translocating it to where that species is abundant, it gives that person a warm feeling. Compassionate stewardship might more wisely, however, dictate humanely taking its life because such displaced mammals usually search in vain for their original home and have little chance of finding a vacant niche and mate. They probably were forced out because they were surplus. The released mammals will cause additional intraspecific fighting before their almost inevitable early death. Research on released displaced mammals shows a high mortality.

Relocation of wildlife species into unoccupied suitable habitats by wildlife professionals, after appropriate analysis and with appropriate care and handling, has proven highly successful and remains a basic and essential technique for wildlife restoration. However, being released in an unfamiliar environment causes much stress and many animals die. In the rare situation where a displaced wild animal is relocated to an occupied environment and survives even though a prior analysis of habitat and population status was not done, its continued survival will still be challenged daily because it must compete with others for food and space. Releasing unwanted, displaced mammals should be prohibited by law unless reintroducing a native species to reestablish viable populations, after appropriate analysis indicates that favorable unoccupied habitat exists and where appropriate care and handling of the animals can be provided.

NATURE CAN BE TENDER AND DELICATE BUT ALSO HARSH AND CRUEL.

Wildlife live off other organisms and, in turn, are eventually eaten. Wild animals, unlike pets and domestic livestock, must be constantly vigilant to avoid being injured or killed by other animals. In nature, wild animals rarely die suddenly or without suffering. Nature has no life-support devices or homes for the elderly. Most small rodents can easily live 5 to 10 years in a laboratory, whereas with nature's food web few attain one year of age. This is why many species must produce large numbers of offspring.

Nature is not composed of species compassionate to one another; it is a battlefield where often bizarre types of cruelty are inflicted on animals. Most predators are not humane in their killing methods. With my students on several occasions we have watched coyotes (Canis latrans) partially consume sheep that were still alive. To avoid being repeatedly attacked in the neck, the sheep lies silent and motionless while the

Nature

coyote consumes its small intestine, the part of sheep many coyotes seem to relish. Predators may play with injured prey or use them to teach their offspring how to kill, as cats often do with captured birds and mice. In Argentina I learned that a mountain lion (<u>Felis</u> <u>concolor</u>) often killed as many as 15 or more sheep in one night when training her cubs how to kill, but carried away only one for food. The same surplus killing of goats by lions has also been observed in California.

NATURE IS BEAUTIFUL, BUT SURVIVAL OF THE FITTEST CAN BE BRUTAL.

Every organism in nature, humans included, considers itself number one and fights tenaciously, even unconsciously, for its right to live. **IF NATURE IS NOT GUILTY FOR HER BRUTALITY, THEN DEATH BY NATURAL AND HUMAN PREDATORS CAN BE CONSIDERED PART OF THE WHOLENESS OF LIFE.** Humans, as predators, operating under regulations, can play a vital role in the harmonious functioning of an animal community even though some wildlife is killed in the process. Remember, to maintain her balance, nature must have a mortality factor, including <u>meat</u> <u>eaters</u>.

It is frequently stated that humans have upset the balance of nature. What is implied when we say nature is no longer in balance? Is it bad for us to change nature? Bad for whom or what? **HUMANS DO NOT ELIMINATE NATU-**

Animal Rights Vs. Nature

RAL BALANCES, THEY MERELY INITIATE NEW BALANCES. That is why we must carefully manage. Whether these balances are better or worse depends upon one's viewpoint.

People cannot make "advancements" in this world without gaining some measure of control over nature and manipulating its balance to their advantage. How to utilize natural resources, as people must, but preserve nature's balance is not easy in a world dominated by economics that require growth for progress. **THE CORRECTNESS OF ENVIRONMENTAL CHANGES, i.e., HABITAT ALTERATIONS, DEPENDS UPON ONE'S VIEWPOINT.** People's different philosophies on this subject appear to be whether short-term population upheavals are going to occur or whether long-term, more "natural" (original) stability should prevail. What was present originally is generally considered "in balance." A basic question is what type of "balance" should be our goal, and what role should man have in both adjusting that balance and in controlling perturbations that may alter the original balance, as often happens when species are overprotected, for example, in national parks. The public has a moral obligation to manage and control wildlife in environments they have occupied.

If all modifications of the environment are assumed to be wrong on moral or ethical grounds, is it wrong for people to exist? Whether our environmental disturbances are good or bad is usually a personal judgment rather than a moral or ethical one. How people intentionally treat animals is a dif-

Nature

ferent issue and will be discussed later. In modified environments, nature needs our help.

"Nature knows best" is a myth. **IN MODIFIED ENVIRONMENTS, NATURE'S NEW BALANCE MAY NOT BE WHAT WE WANT; HENCE, WILDLIFE USUALLY SHOULD BE MANAGED OR CONTROLLED.** If we utilize our knowledge of the balance of nature's processes judiciously, there is no question that we can create desirable and economically feasible communities and even more favorable ecosystems in terms of human preferences.

City people usually will not tolerate native mammals destroying their home landscaping. Likewise, farmers and foresters object if mammals destroy their crops. Nature knows best only in true natural areas that are not occupied by people, because then whatever nature does is usually not considered objectionable. After Homo sapiens moves in, it is usually unwise to let nature take its course. The view that we should let nature take its course in modified environments from an ethical sense is unsound, for nature is amoral and often even immoral from a human point of view. As H. Rolston points out, ethical conduct is based on human culture, not natural events, and the value of following nature lies in helping us fit in with the natural environment.

Most of us would agree that the symbiotic interplay between man and nature has generated many more diversified and interesting ecosystems than occur in some wilderness areas (a matter of taste, of course). The beauty of home gardens and city parks is due largely to exotic plants, not native species, many of which then become "weeds."

It would appear that the public has been so indoctrinated with unsound balance of nature claims concerning animal rights, humane issues, hunting and fishing, and animal pest control operations that the reality of the processes of the balance of nature in disturbed environments has been obscured. The potential opportunities for environmental compatibility in the new habitats created by man often have been disregarded. We overlook the fact that nature's perturbation of the environment often is neither that which man nor many wildlife species would consider desirable.

IF NATURAL ANIMALS AND HABITATS ARE ALWAYS BEST, MUST WE ABANDON OUR PETS, LIVESTOCK, LANDSCAPING, AND AGRICULTURE?

Because preventing the destruction of habitats is the most important way of helping existing wildlife, people often think that any unnatural modification is heresy. However, the introduction of exotic plants or animals often creates new but favorable substitute habitats for desirable wildlife species, greatly increasing species diversity. For instance, in the Sacramento Valley where I live, home landscaping with exotic plants has greatly enriched the variety of birds that nest or

Nature

spend the winter here or stop briefly during migration. Buildings and bridges provide nesting sites for swallows and the artificial irrigation of rice provides them feeding areas nearby. Due to the improved food supply, robins, hummingbirds, mourning doves, and others now nest where it was once grassland.

Wilderness areas and national parks are places where our objective is to preserve as nearly as possible the original environmental conditions. Few will quarrel with such action. But, when out of necessity, people disturb the environment, they should establish new ecological goals and devise well-planned management and control schemes to achieve populations of plants and animals that will then produce a new but "artificial" balance that is both desirable and harmonious. There is a movement in this direction by those who recognize the field of ecology of modified environments! They recognize that people are present (5,300,000,000 of them) and must be accommodated.

ANIMALS ARE FELLOW MEMBERS OF THE ENVIRONMENT.

People often forget that they are part of nature, and that it is essential that they maintain better harmony with wildlife and other natural resources. But who can moralize for others on the tactics used to obtain supreme enjoyment from wildlife in a home garden?

Animal Rights Vs. Nature

An acceptable wildlife ethic is needed so that we will never be allowed to forget that we are part of nature. We cannot afford to ignore nature because we are intrinsically part of nature. We are not short of ecologists who know how to measure what is happening in natural environments, but we need applied ecologists who can predict the cause-and-effect relationships between wildlife and man-modified environments. Some form of artificial control or manipulation of wildlife species is an important conservation tool that is necessary to protect a species from destroying itself in situations where humans have appreciably modified the environment. We can't discontinue agriculture, although there is room for improvement to reduce the dependency upon chemicals, fertilizers, and the high use of fossil fuels.

It would be interesting to prepare an environmental impact statement about the ramifications concerning any planned increase in urban wildlife. The average urban citizen can hardly be classified as a fanatic nature lover. If a homeowner can't make money, win a prize, or get personal satisfaction from a plant or animal, he or she is likely to consider it a weed or pest in the yard. With prevailing attitudes, wildlife has a dim future in society's urban sprawl because people are preoccupied with benefit/cost ratios and material amenities, and wildlife usually requires the sharing of a resource. To many, wildlife other than on television may be a liability. Birds, at least, are often an exception, although they too may be pests.

Our primary goal ought to be "to achieve maximum

coexistence among all forms of life." But most wildlife need water, and water brings mosquitoes; and the wastes of wildlife and their dead bodies encourage flies. Another basic problem is that wildlife do not recognize property boundaries and many species of predators are wide-ranging. Few animals have small enough territories or home ranges so that they will remain in just one backyard.

ARE PEOPLE HYPOCRITES?

We are so entrenched with thinking we are competing with nature -- the man rules beast philosophy -- that we cannot even bring ourselves to plant part of our gardens for the wildlife we have displaced. Instead, we let ourselves be guided by aesthetics, conventional wisdom, and conformity. The urbanite's niche has become a concrete and plastic domain, with about the only reference to wildlife being that of calling a pest control firm to remove a skunk from under the house, a squirrel that is tearing a hole in the roof, or rid a garage of mice. People's physical environment is usually molded to fit their economic and social requirements. They isolate and insulate their biological territory on the basis of their selfish whims with little, if any, consideration of nature or wildlife. Homeowners have little patience with wild animals that become a pest to them.

When we build a home and establish a garden, we purposely, or unknowingly, displace nearly all of the native

Animal Rights Vs. Nature

species of wildlife and plants that had prior claim to that piece of land. Humans rarely try to favor wildlife with appropriate landscape plantings. Instead, they surround their homes with plants that are hardy, disease- or insect-resistant, easy to care for, and all of this for aesthetic, not economic or ecological reasons. How can you favor insectivorous birds if you kill off the insects? Yet some of these same individuals think that the farmer should not be allowed to reduce a wildlife species that has become abnormally abundant because of the crop planted, even though the species may be causing acute <u>economic damage</u> to his or her basic livelihood. In most instances, such species have increased beyond their normal density as a consequence of land being used to provide us with food, fiber, and other resource needs. Examples are voles (<u>Microtus</u> spp.), ground squirrels (<u>Spermophilus</u> spp.), pocket gophers (<u>Thomomys</u> spp.), starlings (<u>Sturnus vulgaris</u>) and deer (<u>Cervus</u> spp.). There are many more.

WE HAVE A MORAL OBLIGATION TO MANAGE NATURE AS BEST WE CAN ONCE WE HAVE DISRUPTED IT.

To protect endangered species, to manage wildlife, and to control problem animals in ecosystems we have altered and to do so in the most environmentally compatible manner possible, it is essential to thoroughly understand the role of the balance of nature in such disturbed environments. Too often many people in academia and the general public oversimplify

Nature

such issues and advocate "let nature take its course," as if nature knows best. According to R. Dubos, we would not have peat, coal, oil, shale, and guano deposits if nature had not failed in the recycling process.

When species populations are declining in an altered environment, the question can be asked: "Is it better to manage the species or let nature take its course at the expense of the population?" Ernest Thompson Seton recognized that the life of wild animals always has a tragic ending. I believe because of our presence and because nature clearly does not know best, we have a moral responsibility to do what we can to maintain healthy populations of animals and preserve biological diversity. When various groups and individuals object to any proposed management of wildlife, they frequently fall back on their favorite cliche that such action is wrong because it is not natural. One might respond that, because people are natural, isn't what they do then okay? Is it better to control a species that becomes so excessively abundant in a human-modified environment that it threatens other species or should we let nature take its course, even though we cannot reestablish the species' original habitat? Should we let coyotes destroy most or all of the young of the endangered whooping crane, as they surely would, if not controlled?

Because we must modify the environment to survive, why do we so frequently charge that environmental changes made by others have critically upset the balance of nature? Why is it that people often apply the balance of nature concept for others but not themselves? Grizzly bears (Ursus arctos horribilis), now extinct in the Sacramento Valley of California, were once common. I have never met someone who wished these carnivores roamed about his or her home and garden, although many want them reintroduced simply because originally they were there.

SHOULD WE ALLOW WILDLIFE FREE ACCESS TO OUR LAWNS, GARDENS, AND SHRUBBERY, AND PERMIT RATS IN OUR GARAGES AND BATS IN OUR ATTICS?

We hear a lot about what we should do or not do to wild animals; but since we exist and are very much a part of nature, what role or rights do you think we have in the natural scheme of things? Is it wrong, for example, that desert-adapted species prosper when civilization converts forests to deserts? Because wild animals frequently show little compassion to other animals, even of their own kind, doesn't this mean that nature is a life-and-death struggle? If so, what is our role with this survival of the fittest where we have modified the environment and altered the original balance? Should we let a new balance develop "naturally" or attempt to help the species we consider most desirable?

The so-called "natural" order of nature no longer exists where humans have settled, for the original naturally evolved environment has been changed. Is this good or bad? Can you think of examples where you would consider it desirable for

Nature

people to intentionally upset or otherwise alter the balance of nature? Because tornadoes and hurricanes are "natural" and affect wildlife, do you think it wrong of meteorologists to research ways of deactivating them? Because bird feeders are not natural, do you oppose their use?

Whether or not you agree with R. M. Case that man is a natural component of the balance of nature, the obvious problem is that man as a species has been too successful. There are many examples in the world where the human race has tragically upset the balance of nature. Why? One answer is that all environmental problems today are the result of advances in education, science, and technology. However, the principal cause is the high degree of death control achieved during the medical and public health revolution of the past century. Thus, although overpopulation may result in a premature death, the resources consumed by that young person before dying were not available to another person. The same is true with wildlife.

It is interesting how the human race has developed the philosophy that not only is it obscene and illegal for a human to die voluntarily but that we are interfering with the balance of nature by considering it unnatural and undesirable for wildlife to die. **SUCH AN ANIMAL RIGHT-TO-LIFE PHILOSOPHY HELPS TO UPSET THE BALANCE OF NATURE AND CAUSES ADDITIONAL SUFFERING AMONG WILD ANIMALS WHEN ATTEMPTS ARE MADE TO DELAY DEATH AND PREVENT ANIMALS FROM DYING OF NATURAL CAUSES.** It is a well-known theorem in wildlife management that you cannot stockpile wildlife. Do we want the equivalent of human life-support systems for wildlife, most of which are much more prolific than humans?

NATURE'S MORTALITY FACTORS ARE PREDATION, STARVATION, DISEASE, CANNIBALISM, TERRITORIALITY, ETC.; EUTHANASIA OR HUMANE SLAUGHTER ARE NOT AVAILABLE.

When predators maintain a natural (though involuntary) balance with their prey by thinning the population, the surviving prey then does not have their own numbers self-limited through starvation, disease, and other vicissitudes of life; hence, they will be in better physical condition and have a higher quality of life than when natural predators have been removed. One of nature's ways of limiting those populations that have exceeded the carrying capacity is with outbreaks of plague, rabies, or other disease, if starvation hasn't been the lethal factor. Also, we know about the high mortality rate of young when parents are facing starvation. A mortality factor of some kind is also essential to eliminate the unfit to preserve a species competitiveness for survival.

Natural populations of wild vertebrates in relatively undisturbed communities do fluctuate in numbers, but are for the most part quite stable, i.e., they are tough and not delicate-

Nature

ly balanced. They also have a surprising resilience to disturbances by humans. The so-called "web of life" that holds an ecosystem together is not as sensitive to disturbances by man as many proclaim, at least as far as vertebrates are concerned. For example, to artificially alter the density of one species of vertebrate may have little or no effect on the other species of vertebrates present in that community unless one is dealing with a direct predator-prey or competitive relationship. The high degree of adaptability and resilience to climatic and some people-caused decimating factors of all the successful species of organisms in a natural community is apparently due to both extrinsic and intrinsic factors, to be discussed later.

For the most part, each vertebrate species in natural environments responds to environmental forces quite independently of other species of vertebrates, except for local situations, e.g., the effect beaver dams may have on trout. The population density of vertebrates will vary within relatively narrow limits in any particular habitat. This ignores, of course, rare obligate predators and the occasional competition that may occur between different species for space and/or food. Instead of a day-to-day affair, the mechanisms that deal with competition and other population-limiting factors are largely an evolutionary event, i.e., the unsuccessful competitors were long ago eliminated. However, vertebrate populations seldom increase beyond the innate carrying capacity of a natural community, even if all seeming needs of the species are provided artificially in what otherwise are natural habitats. This is prevented by various self-limiting factors common to all species.

Animal Rights Vs. Nature

It comes as a surprise to most people to discover just how independent of each other the different vertebrate species are in a community. Humans, for example, are quite independent of other species, and their viability on this earth would be little affected even if many species were tragically exterminated. If <u>all</u> vertebrates were removed, organisms that must feed on vertebrates to live, and perhaps some plants that require birds or bats for pollination are about all that would be affected. Of course no one wants to lose any species because of their aesthetic values. The point I'm emphasizing is that the reason we want to protect viable populations of all fauna and flora is primarily for our own ethical and moral reasons, not just because we think it's necessary to save our lives unless carried to the extreme.

Mankind can modify ecosystems a great deal without seriously affecting the basic flow of materials and energy in the ecosystem, thus not destroying the vitality of the biotic communities within these ecosystems. But major alterations over large areas, e.g., the destruction of most tropical and other forests, would be serious to us. The potential greenhouse warming effect from the increase in CO_2, acid rain, water pollution, etc., need to be reckoned with. However, as long as the earth has abundant producers such as green plants, the survival of humans will be largely unaffected, even by rather drastic changes in the species composition of most consumer organisms, whether native or exotic. We benefit when food and resource production is increased but suffer when we end up producing deserts. The basic functions of resource recycling is necessary to preserve the earth's vitality.

Nature

Fortunately, due to our conservation ethic and other values, we do want to preserve as many original habitats as possible.

ANIMALS ENRICH OUR SPIRIT.

Animals enrich our spirit, whether they are wild, pets, or livestock, and the Endangered Species Act has done much to help justify why all types of animals should be preserved. Animals deserve respect as fellow members of this planet, but it is natural that people favor those they can use. All life forms have inherent worth, and we do not live independent of other forms of life. Even though people's values differ, we should strive to share similar ethical concerns about nature and try to leave a healthier environment to future generations. Yes, we need ways to sharpen people's perspectives of environmental issues. It is not easy. Once a specific philosophy has become deeply entrenched, it is very difficult to overcome the public's resistance to a change.

To show how independent vertebrate species are of each other, let's consider deer, the most significant wild herbivore of North America. If all the deer of North America were suddenly removed, the effect on all other species of vertebrates would hardly be measurable for several years, except perhaps for such predators as the mountain lion, wolf, and coyote, and species-specific parasites. The main environmental disturbances caused by the disappearance of these dominant herbivores would be a modification of the habitat due to the lack of their

Animal Rights Vs. Nature

browsing and grazing. Perhaps after a few years, without deer grazing or browsing, significant changes would occur in the composition and structure of the vegetation, and this alteration of the habitat might then affect other species including small birds. Similarly, it would be hard to measure the effect in natural communities that would result should all of any particular species, such as robins or jackrabbits, be removed from North America. A new balance would develop.

A number of biological factors are responsible for determining the stability and the balance of nature. Perhaps one of the most significant factors for animals is the suitability of the habitat. Animals will do well if they are able to immigrate or be introduced into a locality where the environment provides suitable habitats for them. If they get there but the environment is not satisfactory, they, of course, will not do well. Therefore, habitat preservation is the number one priority for preserving species, not the protection of individual animals.

Just how animals are adapted for different types of habitat conditions is still largely unknown. We do not know the inherited behavioral traits that seem to favor the successful existence of an animal under certain types of habitat conditions. This area needs much more investigation. It is known that most animal populations self-regulate. If all the environmental factors needed are present, what is going to keep the animals from continuing to increase beyond the carrying capacity and result in the destruction of their habitat? Limiting factors in nature include many intrinsic and social factors

Nature

including variable reproductive roles, various types of stresses, food supplies, diseases, territoriality, and aggression. If all aspects of the environment are favorable to the survival of a species, and natural interspecific mortality factors eliminated, the ultimate break in their population growth comes from members of that species, their own social self-limitation behavior. Changes in the conditions of the habitat merely raise or lower the upper density limits that are controlled by these regulating feedback mechanisms. However, animals such as deer will destroy their habitat if the efficiency of natural predators to control their numbers is altered. The deer population is then regulated by the food supply and disease.

Animal movements are important factors affecting population densities. These include dispersals; that is, emigrating from the population or being an immigrant into another population or locality. Dispersal is the way territorial species, e.g., carnivores, control their densities. Weather, fire, and other environmental perturbations also affect the carrying capacity of populations. Food is usually not too significant unless there isn't enough food. When we add more food to the environment, we seldom see populations increasing much more than they would attain naturally in the most favorable habitats, except with many herbivores.

Many people think that predation is the main factor regulating population densities. Natural predation is very important in maintaining the balance of animal-plant communities. However, natural predators are not very effective in greatly depressing the densities of their natural prey species.

Animal Rights Vs. Nature

In fact, the carrying capacity of a population of prey species can be maintained at a higher level over time with the presence of predators than could exist if the predators were removed. When these predators are removed, their prey species usually responds with an increase in population growth, but it is only a temporary increase.

When their predators are removed, deer and other herbivores tend to overpopulate and then overgraze and overbrowse, and this overutilization of their food supply is then the main population-limiting (mortality) effect due to the loss of predators. Such overutilization of the vegetation will result in a less productive plant-animal system. Consequently, with the presence of wolves, mountain lions, coyotes and bears, deer numbers can be sustained at a higher density than if there were no predation by wild animals or man. But where humans also prey on populations of deer, it is often necessary to control the deer's natural predators due to their competition with man. Examples are the wolf and moose in Alaska and, currently, mountain lions and deer in California.

When a predator becomes too abundant, i.e., gets out of balance, it may need to be controlled to prevent undesirable overutilization of their prey from occurring. With large carnivores, prey can be shared by the predators and humans. Otherwise, intentional predator control is seldom beneficial to wildlife, except to protect endangered species or other species that have become more vulnerable prey due to habitat alterations. It is often essential to control predators when wildlife are being reintroduced into a favorable but unoccupied habitat.

Nature

The artificial introduction of vertebrate predators to control pest prey species also is seldom justified or effective. This biological control approach is usually successful only on small islands where the introduced predator is not expected to remain.

One reason coyotes and other livestock predators become troublesome is because humans provide them with prey species, e.g., sheep and cattle, which, in the process of domestication, have lost most of their predatory defenses. Unlike grizzly bears, wolves (<u>Canis lupus</u>), or bison (<u>Bison bison</u>), which cannot be tolerated in areas where many people live, coyotes can and do live close to large human populations, which creates problems. They even live in cities and parks where they feed on garbage, cats, small dogs, fruits, and attacks on children are not uncommon.

It is frequently erroneously stated that there are few, if any, empty niches in nature, and that an exotic plant or animal can be introduced only at the price of a species already present. That a native species will always be displaced does not seem to be the case, for most countries today have a much more diverse fauna and flora than would be present naturally, due to introductions. However, I do not wish to imply that introductions can be made with impunity.

The biological price of introductions may be very high if the species of vertebrate being introduced has no near relative already present so that a natural stability has not already evolved between that type of animal and the plants that are present. New Zealand is a good example. Since its unique endemic vegetation evolved in the absence of any grazing or browsing mammals, environmental instability resulted when Europeans introduced European rabbits (<u>Oryctolagus</u> <u>cuniculus</u>), hares (<u>Lepus</u> <u>europaeus</u>), many species of deer and other mammals, and domestic goats, pigs, and other animals became feral. As a consequence, in New Zealand today the species composition and density of plants in many of their national parks have been permanently altered. A few sensitive plants, including some large species of trees, have been completely eliminated in many of the national forests and parks. Yet, if most of these same animals were introduced into the temperate regions of the United States, very little instability would result because we have many other species of animals already present that are quite similar to those introduced in New Zealand. There were no mammalian predators in New Zealand. In fact, the only land mammals were two kinds of bats. The many genera of extinct moas were browsers and grazers, so the vegetation was adapted to these giant birds.

Forests and wildlife are renewable resources but, of course, natural forests and other natural environments cannot at the same time be utilized by us and still be kept in their exact natural state. Any use by humans tends to modify the environment, and it takes very little change in habitat conditions to affect various species of vertebrate wildlife. Usually when the environment is modified, most of the vertebrate species tend to do less well than they did in the original habitat even though it may appear that there is more food and cover for the species than before. Some species seem to be able to adapt readily to new environmental conditions, although most of them cannot. We don't know all of the factors involved. For example, in western desert conditions, we usually find the greatest variety and density of small rodents. These rodents can exist there only because they have a good food supply, and plants are the basis of their food web. Yet, the desert is where you find the fewest plants. If people bring moisture into these deserts and produce perhaps a thousand times as much of the primary producers -- the plants -- and even though the new plants may be nutritious to the various species of rodents, we find that with this increased production of vegetation many of the rodent species tend to disappear. It seems they are adapted to certain specific habitat conditions rather than just to natural predators, food supply, and breeding places.

All wild animals in nature, whether coyotes, eagles, rodents, or rabbits, play some environmental role. However, even though these functions may not be really essential for the welfare of people or the benefit of most other species of wildlife, we consider their role highly desirable.

I like to assume that wild vertebrate animals think and learn by experience. However, we know that many of their behavioral traits are genetically based; for example, we have shown that coyotes inherit the trait of attacking fleeing sheep by the neck. They do not have to be taught such behavior; they are born with this skill, whereas those in the cat family train their offspring how to be a predator.

Animal Rights Vs. Nature

With few exceptions, coyotes, rodents, birds, and other wildlife species are not like links in a chain where the loss of one species will have deleterious effects on all the others. **WHAT IS RIGHT OR WRONG ABOUT MOST BALANCE OF NATURE AND ANIMAL WELFARE ISSUES DEPENDS UPON HOW YOU DEFEND YOUR VIEWPOINT.**

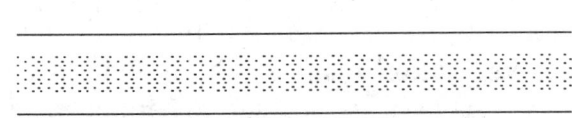

The following questions are to enable you to test your own objectivity about these matters.

1. Because the balance of nature requires a fairly high premature death rate for each species, do you believe that artificial life-support systems for animals that are destined to die will undesirably upset the natural balance?

2. We hear a lot about what we should not do to wild animals, but because we exist and are very much a part of nature, what environmental role and (animal) rights do you think we have in the natural scheme of things?

3. Even though, hopefully, you oppose desertification,

Nature

have you ever thought about how much civilization's conversion of vegetated areas to deserts benefits the desert-adapted species?

4. If someone prevents a "natural" death of a wild animal, do you consider such an act as an undesirable disruption of the balance of nature?

5. Can you be objective and give examples where <u>you</u> would consider it desirable for people to intentionally upset or otherwise alter the balance of nature, other than your house and garden?

6. Do you oppose the "death ethic" for wildlife that encourages people to assist nature by preying on the "surplus" individuals for the well-being of populations of animals?

7. When wild animals die naturally, do you think they die nicely?

8. Because wild animals frequently show little compassion to other animals, especially their own kind when the population density is high, doesn't this mean that nature is truly a life-and-death struggle? If so, what is man's role with nature's survival of the fittest, especially if he has modified the environment and upset the original predator-prey balance?

9. Should we assist nature by acting as a predator to help

Animal Rights Vs. Nature

maintain a healthy balance within a prey species when it is no longer possible for us to reestablish the original predators in sufficient numbers?

10. Because all animals are physiologically and evolutionarily predisposed to overproduce, do you agree that a basic law of nature is that there <u>must</u> be a high premature (before reproduction) death rate before any wildlife species can exist over time in healthy populations?

11. Are we justified in destroying individual animals when it is clear that such action will benefit many individuals within the population of that species, or even with another species?

12. Do you agree that the life span of captive animals is usually much longer than wild individuals that must daily face the adversities of climatic extremes, starvation, disease. and danger from other animals?

13. Despite the unfortunate human-caused extinctions of certain fauna, do you nevertheless agree that all land masses, except the two poles, today have a richer, more diverse fauna (and flora) than would be present naturally because of the many introduced exotic domesticated and wild species? Do you think this is partly good or all bad?

14. Isn't the so-called "harmony" or "balance" in nature really no more than the current degree of biological

Nature

equilibrium established by a desperate struggle of each organism to survive?

15. Because the population densities of lemmings, arctic hares, and their predators fluctuate a great deal naturally, even where man has not altered the environment, do you consider nature as being in balance in these regions?

16. Nature fosters the most bizarre types of cruelty. Are you in favor of all of it because it is "natural"?

17. Do you think animal suffering is all right when it can be classified as "natural"? Do you consider all activities by humans that affect wildlife "unnatural" and thus bad?

18. When it can be done easily, do you think it is ethical to offer assistance to animals that are suffering "natural" brutality? How about if the suffering is a consequence of our altering the environment?

19. In disturbed ecosystems where natural predators have been permanently reduced, or where a prey species can now multiply beyond any possible control by natural predators, is it better to let all of the animals regulate (self-limit) their own numbers, usually with much suffering, or for humans to control or harvest the surplus animals so that much less overall suffering will occur?

Animal Rights Vs. Nature

20. Because tornadoes and hurricanes are "natural," is it wrong for meteorologists to try to find ways of deactivating them?

21. As bird feeders are not "natural," do you oppose their use?

22. Are people justified in converting native "weeds" into what they consider a more attractive garden or park composed of exotic plants? In California, was it a mistake to introduce the ring-necked pheasant and striped bass?

To conserve gannets (<u>Morus</u> <u>serrator</u>) at Cape Kidnappers in New Zealand, their nesting rookery on this high cliff top plateau now has to be protected from predators that have been introduced. Originally New Zealand had no land mammals except two species of bats. Sea birds usually nest on cliffs or islands to escape mammalian predators. (W.E. Howard)

Zoos provide the only opportunity most people ever have of seeing the major wildlife species of the world. Animals born in a zoo must be given special protection if released into the wild. The author has experienced tigers (<u>Panthera</u> <u>tigris</u>) in Burma, India and Malaysia, but none provided the close view of this great carnivore as this animal in Peking Zoo, Beijing, China. (W.E. Howard)

3. ANIMAL RIGHTS

Much confusion prevails if one tries to analyze some of the current animal welfare and animal rights issues. There has been a dramatic shift of people's philosophy about animal welfare to one of animal rights. Just what do we mean by the terms "animal rights" and "animal welfare"? How do the morals and ethics of these terms apply to wildlife and confined animals we use in various ways? How do these terms relate to human beings? Do we need to apply these terms to all our own direct relationships with animals? Certainly most people agree that we must deal with these issues in a more socially responsible manner than has sometimes been done in the past, or is currently being done in certain situations. When we see animals suffering, do we unconsciously compare it to a person being similarly treated, even though we are not granting such vested rights to animals? I agree with one of my reviewers of a related article that just as human rights are building blocks for human-human morality/ethics, vesture of animal rights logically leads to human-animal morality/ethics.

Even though I question animal rightists' views, I am grateful for one positive value of the recent animal rights movement, and that is the increased consciousness people now

Animal Rights Vs. Nature

have about the welfare of pets, livestock, laboratory animals, and wildlife species. This is good and timely. But this philosophical change has been coupled with a misunderstanding of nature and animal suffering. Most of the lay supporters of the animal rights movement, even though grossly misled about the laws of nature due to the emotionalism so rampant these days, are honestly and, of course, in part justly concerned about animal welfare. In contrast to animal rights, animal welfare is an expression of kindness and concern for the well-being of animals that people use in the field, at home, or in the laboratory. This is a philosophy that I strongly endorse.

There has always been a need for welfare organizations doing their best to prevent unwarranted suffering of animals, but the increased radicalization by animal rightists is not constructive. Animal rightists, depending on the group, oppose any management, use, or exploitation of animals, and, for animals known to feel pain and suffering, they want to give them the same legal status now afforded people. Animal rightists give animals legal rights equal to those of people, yet fail to recognize that nature is based on survival of the fittest and displays no equal rights between individuals or species. Why try to give animals rights which they don't acknowledge as proper, nor recognize with each other, and would certainly violate these rights if they had them?

Leaders of extreme animal rights organizations want to stop any and all exploitation of animals regardless of how painless and humane the use of an animal may be. They do

Animal Rights

not think human beings have a right to knowledge gained from experimentation on animals. The February 1990 issue of The Washingtonian says that nearly 90 research laboratories have been raided in the decade-old animal rights movement. Close to my home, in 1987 $4,700,000 of fire damage was done to a veterinary diagnostic laboratory under construction at the University of California, Davis. Activist animal rightist students explained to me that the torching was justified because "animals should not be domesticated." Similarly, when a nearby livestock auction yard was burned, they said it was because "some of the livestock had been pastured on public grazing land," a multiple land-use practice they think is wrong. For such fires the animal rights leaders say economic loss is the only thing such animal exploiters understand.

The animal rights debate became more radical with the publication of the book "Animal Liberation" in 1976 by Dr. Peter Singer of Monash University, Melbourne. Animal rightists are showing that exploitation of animals cannot be promulgated based on ethics when the decision is going to be political, not ethical. Animal rights, which started as an ethical issue, has now become a political issue. Laws, not the users of animals, are now setting the rules. And it is the media that control the dissemination of information.

In 1990 I attended an enlightening university conference on "Animal Rights and Our Human Relationship to the Biosphere." Some of the key speakers, and some of the university students, preferred to use it as an animal rights activist rally. I object to having them dictating the ethics of

Animal Rights Vs. Nature

others. With them, objectivity and constructive discussion were out. Animal rightists tend to convert any public debate on the subject into a rally, probably because it is impossible for them to demonstrate credibility in an objective discussion of animal rights. Instead, they try to inflict their ethics on others. But ethics are not black and white; there are many gray areas. A highly respected scientist was invited to present a pro paper on the use of animals in biomedical research and education. Perhaps because it was presented too convincingly, the verbal attack against him was disgusting. The audience wanted a rally, not constructive deliberations.

Many questions can be asked of extreme animal rightists. Is rodent-proofing a house denying rodents their rights? Should we share our woolen clothing with clothes moths and tolerate rats, fleas and bedbugs in our houses? The animal rightists I know still kill ants that invade their kitchens. Are rats equal to people, and what type of respect should we give a dead rat? Do animal rightists offer their bodies to science to spare laboratory animals? How do they view quality of life of animals with respect to merciful killing of animals? Should a horse be kept alive after a train has severed both its hind legs?

Are we really trying to be humane to individual animals and species populations when we attempt to reduce the suffering of individual animals by trying to prevent their "natural" deaths? When raccoons, skunks, coyotes, and other animals are not wanted in an urban area, or have ventured into habitats unfavorable for them, what do we do? Turn them over to the local humane society for proper disposal by euthanasia? Unfortunately, this is not usually the case. Most people seem to want such misplaced animals caught and then released somewhere in the wild "to give them another chance." Why not? At first it sounds logical!

How can one equate the suffering of animals that are used in farming or research and those shot, trapped or poisoned, with that of natural suffering, e.g., a slow death from starvation and disease, cannibalism, intraspecific battles, or being maimed, killed, or eaten alive by a predator? One cannot dismiss all of the latter forms of suffering by saying they are natural cruelty if the causative factors are due, even in part, to our upsetting the balance by modifying the environment. We should assist nature, especially when such people-caused suffering is largely controllable with modern wildlife management techniques. In human-altered environments, sportspersons, subsistence hunters, and animal control personnel who may artificially limit animal numbers, can greatly assist in preserving a healthier balance among animal populations and their habitat resources, even providing protection and sustainability of threatened or endangered species and their habitat.

The animal rightists' philosophy seems to ignore that pain and suffering are important factors of evolution and are required for certain segments of the natural system to function. We all know what pain is to ourselves but fully understanding what pain and suffering are to animals still largely eludes scientific description and are often difficult to verify. More information is needed. Since animals often voluntarily do things that inflict pain, it is not easy to evaluate suffering anthropomorphically. Before an animal can "suffer," it must have awareness and be distressed emotionally, hormonally, or consciously, and have its behavior either subtly or overtly affected.

Pain and suffering are not equivalent, and suffering is a more preferred term over pain. Animal pain is supported by behavioral responses to avoid what is a negative unpleasant sensation. Suffering is an unpleasant emotional response, including anxiety. Even though pain and suffering are not equivalent, they do overlap. The less-developed nervous system in lower forms of animals probably means they do not experience pain and suffering as do sentient animals. The study of animal pain is recent. The first conference was held in 1983. It has been suggested that since animals cannot analyze the factors that cause them pain, this could mean they may even suffer from a painful stimulus more than humans.

The subjective experience of consciousness implies our appreciation of love, beauty, meaning, and values that make up so much of our daily lives. It seems to me that evolution demands that animals respond negatively to unpleasant stimuli (pain and suffering) in order to develop escape mechanisms to ensure their survival. Most animals, certainly vertebrates, quickly learn to retreat and avoid what we ascribe as probably being painful to them.

Even though nature causes and demands pain and

Animal Rights

suffering, this is not a justification for us to inflict unnecessary suffering. It is still incumbent on people to have a compassionate respect for all animal life. Perhaps we must consider whether there can be joy and happiness without contrasting it to pain and suffering.

Animal welfarism is often a struggle between animal lovers and our economic goals. Profitability of domestic animals may be lost if the demands of some groups are met requiring that naturalistic environments be provided to all captive animals. Unfortunately, little is known about pain, stress, and frustration, which may be experienced by captive animals. Animals do not have the same perceptions as humans, and often when they appear to be under stress or duress, they may not actually be suffering. Stress can be quite pleasurable and actually sought voluntarily. Ask any hiker or jogger.

We cannot exclude man from the real world of nature even though human consciousness is markedly different than that of animals and seems outside the realm of natural science.

IT IS DANGEROUS TO ANTHROPOMORPHIZE AND GIVE ANIMALS HUMAN CHARACTERISTICS, SENSITIVENESS, FEELINGS, AND CONSCIOUSNESS.

Even though some animals clearly demonstrate insight

Animal Rights Vs. Nature

and reason and are very much aware of themselves and their life struggles, i.e., show intelligence, these sentient beings seldom demonstrate conscious compassion for other species as do humans. Some feel that "higher" animals, which have entered a sort of domestic partnership with people, should have legal rights equivalent to humans since they are claimed to experience emotions similar to those of people. I consider this invalid anthropocentricity. Just because some animals typically express certain behavioral patterns does not necessarily imply that such behavior is needed for the welfare of the animal; e.g., is it unkind to keep people alive who may not possess all the prowess of physically fit individuals?

A philosophy that emphasizes Albert Schweitzer's reverence for life of all animals may in part be contradictory to the necessities for quality of life. There is no quality of life for a population of animals that has exceeded its carrying capacity and is being limited by species self-limitation factors like disease and starvation. To me the issue is not animal "rights" but ethics that oppose any unnecessary mistreatment or prolonged suffering of animals over which man has direct control. Actually, animal rightists usually focus on individual animals, not the welfare of a population or a species.

If one agrees with Michael Fox that all animals have souls, how does one accept the brutality of nature, which is far more cruel than people's use of animals as pets, beasts of burden, food, or sport? After all, people are part of nature and the continuity of evolution. If you look for examples of one species having true compassion for another, none begins to approach the compassion most people have for each other, which is very unusual in nature. B. Callicot states in "Animal Liberation: A Triangular Affair" that most animals survive by killing other animals and that those who believe that animals have an individual right to life must logically look upon predators as "...merciless, wanton, and incorrigible murderers of their fellow creatures." It is not clear how the humane and animal rights groups can justify their goals for animal populations.

OVERPROTECTION OF ANIMALS INEVITABLY CAUSES SUCH POPULATIONS TO FACE HORRIBLE DEATHS FROM MALNUTRITION AND DISEASE.

The carnage of wildlife on our highways is almost unbelievable, surely ranging somewhere between one-half million to over a million kills a day. Road kills are not very selective, and victims include mountain lions, black bears, very large numbers of deer, rabbits, skunks, opossums, nocturnal and diurnal rodents, large numbers of barn owls, many song birds, young waterfowl in particular, frogs, snakes, turtles, and other creatures. More wary and less frequently killed species are weasels, coyotes, and bobcats. Dogs and cats are vulnerable. Nearly all drivers have experienced the unpleasant trauma of running into or over some form of wildlife. Yet "extremist" groups tend to accept this form of unnatural mortality of wild animals, perhaps because they are part of it or can do little about it.

Animal Rights Vs. Nature

In January 1986 the California Biomedical Research Association reported that a public opinion survey made in August 1985 showed overwhelming public support (79%) for use of animals for health-related research, and a substantial majority were in support of basic research (77%) and 69% opposed organizations trying to ban research on animals. I prefer utilizing "unwanted" cats and dogs for research and teaching instead of breeding them for this purpose. But, of course, when such laboratory use is justified, these animals must be treated in as humane a manner as possible.

It has become fashionable for some groups to criticize zoos, claiming that animals have a right to be free in their natural habitats instead of confined in a zoo, even claiming that the educational and preservational premise of confining scarce or endangered species in zoos is false and self-deceiving. I presume this is because they think animals confined in this way are deprived of all the perils associated with living naturally. The California condor seemed to have little chance of being able to survive in the wild, so why not see if a population for future releases can be raised in zoos? I suggest we stop criticizing zoos and use our energy and funds to save natural habitats. Zoos and habitats have unrelated roles and the most valuable zoo animals are often those whose habitats are scarce or no longer exist. Most zoo animals were born in a zoo and could not survive if released. When people get acquainted with wildlife in zoos, they are more likely to favor saving the animals' natural habitats.

When I suggest that some people anthropomorphize

Animal Rights

when considering an animal's current predicament, I am implying that an animal's awareness or consciousness of a particular situation is not going to be the same as it would be with man. The question of how humans differ from animals -- because we have feelings of right and wrong -- have generated all manner of philosophical debate, with arguments based on morality and ethics concerning the rights of animals and how we should use our power to control the future of all species, according to B.E. March.

With the animal welfare issue, the public needs to examine the other side of nature's coin. **INSTEAD OF INFLICTING GREAT CRUELTY BY TRYING TO SAVE DISPLACED ANIMALS, WHY NOT BE HUMANE AND SHOOT, POISON, KILL-TRAP OR EUTHANIZE THEM** when necessary to establish a healthy population balance?

Some claim that the battery production of chicken eggs is inhumane. In contrast to the life of a jungle fowl, are not laying hens always provided adequate food, fresh water, shelter from the weather, shielded from predators, and protected from diseases and parasites? True, the hens have little space, and this is worthy of discussion, but does that cause suffering of the degree claimed to occur in a domestic breed that has no knowledge of what it may be missing? Are not many strains of cats, dogs and birds also bred for a confined existence in houses?

Animal Rights Vs. Nature

SELF-RIGHTEOUSNESS ABOUT THE SANCTITY OF ANIMAL LIFE CAN BE CAUSE OF GREAT CRUELTY.

"Bambi-ism" can be very cruel to an animal population, and the recent surge of animal rights emotionalism about animal rights has misguided many dedicated and sincere people about the laws of nature. In nature's scheme of things it is often necessary to cause some animals to suffer in order to be humane to an entire population of that species. Also, is it better for an animal to have been born and live longer than its average wild counterpart even if it is going to die prematurely, but humanely?

Interfering with nature's death ethic may make our conscience feel better, but this warm feeling we experience inside may not be from a compassionate act for it may actually cause much animal suffering. Animal rightists need to recognize that a philosophy which emphasizes protection of all organisms in modified environments will often be contradictory to the necessities for quality of life of such animals. There is a great difference between the quality of life of, say, a deer herd whose population does not exceed the carrying capacity of its range, with sickly undernourished deer where their population has exceeded the carrying capacity.

SYMPATHY VS. COMPASSION

When we capture misplaced, unwanted animals and then translocate them to what appears to be a natural habitat, do we analyze the trauma and probable fate to which we destine the released animals? Or do we go through this behavior primarily because it makes us feel good to presumably have helped a wild animal? In reality, we may be causing undue suffering to each released subject which, if it is not quickly dispatched by a predator, is certainly not going to be welcomed by other members of its species.

Instead of translocating misplaced animals, if you have true compassion for the animal, you will either humanely euthanize it or -- if you insist it should live -- offer it to a zoo or animal park, which is not pleasant for a previously free animal. Few translocated animals survive. A Canadian study (Rosatte and Macinnes 1989) concluded that the relocation of urban raccoons (Procyon lotor) to solve nuisance problems is not recommended because of the high mortality and severe weight loss in juveniles and the potential for disease transmission due to their large exploratory movements, when probably trying to find "home." Unless man has been harvesting individuals from that species' population, the chance of a released individual finding a mate and a vacant habitat is slim indeed. Only 15% of deer relocated at great cost from Angel Island, California, survived the first year.

Animal Rights Vs. Nature

How many wild horses and burros can the environment support? This question does not seem to be considered by those groups that insist on capturing, at a ridiculous expense, the surplus animals in order to find another, often unpleasant, home for them. Since the animals overgraze the range and seriously compete with mountain sheep and other native wildlife, why not convert the surplus into pet food? Such recycling is quite natural, establishes a better balance, and is probably more humane than what many of the animals experience after they are adopted.

I find it easy to respect people who are opposed to drilling for offshore oil if they don't drive a car or use energy or products from oil. Similarly, I can admire no-growth enthusiasts if they have no children, or the extreme anti-smokers if they also don't drive an automobile.

There is no point in courts fining polluters only to have the money go into federal, state, and local general funds; rather, they should be used to buy habitat and for habitat repatriation. The replacement value of animals has little value, so polluters should be fined to pay for habitat value, which has a much greater monetary value. Instead of just trying to avoid contaminating with pollutants, we need to think how we can achieve less pollution than there was before. We need to dig ourselves out of the hole, not deeper into it. In the infamous Valdez, Alaska, 1989 oil spill, care of contaminated sea otters cost about $1,000 per day per otter. The public should know this since they pay the bill. The clean-up costs to Exxon Oil Company are tax deductible, hence also done at public expense.

If the human body is exposed to enough lead, death is the result. What is happening to this planet due to oil spills and other pollution is no different. The environment is the victim. The goal of public policy must be to reduce the need, hence hazard, of oil and other pollutants such as plastics and other chemical compounds new to our environment.

We would never consider letting flocks of sheep or herds of cattle become emaciated, sickly, and even starve. But this happens to deer and other wildlife, which we too often overprotect from both natural and human predators. To establish a new balance, we should use the most humane procedures available. When a doctor treats us for an illness, he or she must consider the whole patient. Environmental problems are equally complex.

ONCE PEOPLE HAVE DISRUPTED THE NATURAL BALANCE, WE HAVE A MORAL OBLIGATION TO CORRECT IT.

It is often difficult for professional wildlife managers to comprehend the behavior of some of the extreme animal rights groups. One example in California is the repeated legal blocks used to prevent the removal of feral goats on the Channel Islands. These exotic animals were to be humanely shot to preserve endangered plants and the habitat from severe overgrazing by the goats, when legal blocks were instigated to force their live capture. Another example is the many legal blocks the California Department of Fish and Game received when wardens were going to shoot the surplus deer on Angel Island in San Francisco Bay. Animal welfare groups did not want the deer shot nor to have male coyotes introduced to do it nature's way. Some groups demanded the deer be fed. Others wanted them sterilized. They finally forced some of the emaciated deer to be trapped and released on the mainland, where they promptly died. Overpopulations of deer on the Florida Keys and with the Smithsonian herd are other examples. Animal welfare groups must recognize the necessity for a certain proportion of animal populations to die prematurely, even if by artificial thinning.

NATURE DOES NOT GIVE ANIMALS SPECIAL "RIGHTS" EXCEPT THE RIGHT TO TRY TO SURVIVE.

Many animal rightists claim that all wildlife have as much right as we do to live in peace, even though the animals do not reciprocate in respecting the rights of other species or, for that matter, man's rights. Animal rightists need to explain why they think nature's brutality is better and more humane than the more humane management and control techniques man has available. They do not seem to recognize that premature deaths are essential to a balanced ecosystem. Wildlife rarely attain old age in the wild. In fact, with few excep-

Animal Rights

tions, ANIMALS LIVE MUCH LONGER IN RESEARCH FACILITIES AND ZOOS THAN THEY DO LIVING "FREE." In nature, animals do not have rights to humane slaughter, yet I believe most researchers and animal control workers today try to invoke humaneness in their management and control of populations, and regulations require it. It is not clear how animal rights groups can justify their goals for animal populations when **OVERPROTECTION OF ANIMALS INEVITABLY CAUSES SUCH POPULATIONS TO FACE HORRIBLE DEATHS FROM MALNUTRITION AND DISEASE.**

Wildlife cannot be stockpiled by following the animal rights movement of providing complete protection, for all species produce surplus individuals that must be removed and -- if not taken by people -- will die by the various inhumane ways of nature. Environmental humaneness and animal rights issues need to be assessed on their merits, not from an irrational, sentimental, or emotional base. If it is wrong to prevent coyotes from killing lambs, cats and small dogs, should people stop raising sheep or having pets in such localities rather than control the coyotes? Without control, surplus coyotes would soon move into cities causing even more children to be attacked. Many animal rightists think people should not exploit any animal. They do not recognize that most wild animals exploit each other.

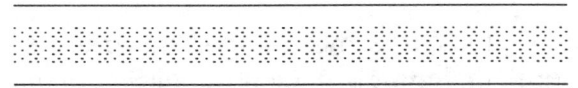

Animal Rights Vs. Nature

Some questions follow to test your objectivity relating to animal rights.

1. Do you think that leash laws for dogs cause more suffering to man's best friend than does caging laying hens. Are you opposed to holding any animal in captivity?

2. Do you think keeping pets is justified? If so, what kinds of pets do you condone? Do you think it is wrong for us to exploit any of our finned, feathered, or furred wildlife as a sport, material resource, or pet?

3. If you claim you are not an abolitionist against all human uses of wildlife, what uses will you tolerate?

4. If you are a vegetarian, do you think it wrong for man to be a predator of wild birds and mammals for food, even when this predation is needed to maintain healthy populations of these species?

5. In your attempts at changing public attitudes and values about humaneness and animal rights, can you give examples of sound reasoning or biological facts upon which to base your arguments other than the few known examples where there has been improper treatment or abuse of animals?

6. Do you think it is more important to use social issues rather than biological facts or basic economics when

Animal Rights

making decisions concerning the welfare of animals?

7. How much of a personal sacrifice are you willing to make to benefit wildlife? Will you change your lifestyle?

8. Are your beliefs concerning animal rights "religious" views, or are you willing to debate the related basic laws of nature and biological principles of the dynamics of wildlife populations? If you are shown to be mistaken in regard to the biological principles of nature, do you think you could change your mind?

9. When love and happiness or quality existence are no longer possible for an animal, is a quick death more humane than lingering suffering, pain and distress in the vicious world of cruelty that is nature's death, or do you prefer that the animal's death be prolonged for as long as nature will permit, as is often done with human life-support systems?

10. If a horse has to have a leg amputated, would you recommend it be shot?

11. Which do you think is more important: the welfare of individual animals, or the health and well-being of their populations?

12. Since nature does not allow many species to live in peaceful coexistence, are humans more compassionate

and humane than wild animals?

13. Do you think you are demonstrating true compassion towards a misplaced individual animal when you capture and release it at another location where you know the species is common, even though it has been proven that most such translocations do not survive, and those that do may first experience many social conflicts with members of their own species before finding a vacant habitat and mate?

14. If you justify translocating misplaced animals because it gives them another chance to live, however small, do you feel it is still justified even if each transplanted animal will cause others of its kind to suffer as well as suffer itself?

15. Do you think animals have the same consciousness, emotions, and anticipation of pain as people?

16. If you support the artificial feeding of starving wildlife, would you change your view if you were shown how such feeding is later going to result in many more animals than you helped survive experience a substantial increase in the intensity of suffering from starvation and disease?

17. If you think all wild animals have as much right as we do to occupy space on this earth, when, if ever, are we justified in taking over the habitat of wild animals or in

Animal Rights

removing them because we do not want them present in our house, garden, or crop?

18. Do you think you have any more of a moral or ethical right to control rats and mice in your house than a farmer does to control animals damaging his crop?

19. How can extreme humane or animal rights groups justify their overprotection goals for animal populations which inevitably will eventually cause such populations to face increased death rates from malnutrition, disease and predation?

20. Do people have a moral obligation to protect from predation animals they have domesticated?

21. Do you think animal suffering is all right when it can be classified as "natural"? Do you consider all activities by humans as "unnatural" and thus bad?

22. Do you think it ethical to offer assistance to animals that are suffering "natural" brutality? Are we justified in destroying individual animals when such action will clearly benefit a population of that species, or another species?

23. Do you agree that captive animals usually live much longer than wild animals that must daily face the adversities of climatic extremes, starvation, disease, and danger from other animals?

Animal Rights Vs. Nature

24. Since self-righteousness about the sanctity of animal life can result in great cruelty to animals because of overprotection and entire populations can be "loved to death," do you agree that proper management and control are much better ways of conserving our wildlife heritage?

It does not have to be natural to help wildlife. Due to a shortage of natural nesting sites, barn owls (Tyto alba) promptly took up residence in the nestbox patterned after ones in Malaysia. It was built and installed in the author's garden by the University of California Raptor Center. Many broods have been raised. (W.E. Howard).

By excluding elk (<u>Cervus elaphus</u>) in 1936, aspen seedlings soon appeared. When the fence around one end of the enclosure was removed in 1938, elk soon killed the exposed aspen. When more of the meadow was enclosed at the other end, aspen sprouts soon appeared. Without the overpopulation of elk, this area would probably have been an aspen forest. Yellowstone National Park, 1950. (W.E. Howard)

4. VESTED INTERESTS

I am not alone. Many conservationists, professional wildlife managers, researchers, livestock operators, and sportsmen share my concerns that the extreme measures pursued by animal rights groups are not helping our wildlife and other animals live happier and more successful lives. Indirectly, such action often causes much more suffering, especially amongst wild animals. We all have our vested interests in the sense that we want to see our own objectives achieved. Certainly livestock producers and agricultural commodity groups lobby to protect their investments from wildlife depredations. But the vested interests referred to in this chapter refer to the "cheap shot / half-truth" technique too often used by extreme groups to raise money, largely for their own financial support, by tricking a well-meaning, highly concerned portion of the public into thinking all their donations will save animal lives and improve wildlife conservation, when often the contributions merely subsidize organization leaders and provide money to finance additional "anti" fires.

Are you getting tired of the deluge of "alert" mail

requesting donations for environmental emergencies? Perhaps the unscrupulous, deceptive solicitations sent out by some groups need to be exposed to save the needed environmental movement from losing credibility. If we don't watch out, the public will no longer believe that legitimate environmental and animal welfare alerts are really genuine. The leaders of these organizations chart the course and steer the ship, but the members don't vote on the decisions. Also, these organizations are desperately competing with each other for the diminishing public dollar. Somehow, new and tougher standards are needed to govern the irresponsible rhetoric so prevalent today in raising tax-exempt funds. Some groups, to help get money <u>for</u> <u>vested</u> <u>interests</u>, have also carried their adversary approach to civil disobedience.

ONE WAY TO RAISE MONEY IS TO POLARIZE AN ISSUE.

One reason many wildlife matters are so difficult to resolve is because environmental obstructionists polarize the issues, then lobby congressmen and agency officials. If they don't get their way, they sue. Since it is difficult to raise these "anti" funds with mediation, instead of compromises, we are deluged with costly litigation. Lawyers go out to win, not compromise, and lawyers have replaced biologists as leaders of many environmental organizations. Former Chief Justice Warren E. Burger (1983) pointed out how law schools have traditionally steeped the students in the adversary tradition

Vested Interests

rather than in other skills of resolving conflicts. Lawyers are natural competitors, and once litigation begins, they use every tactic available to win. Unfortunately, environmental issues are choice litigation material for lawyers. It is clear that we cannot manage animal welfare issues with law enforcement; instead we must have education, especially to neutralize the animal rights propaganda. All I can do with the legal jargon is eavesdrop; I am unable to judge their conclusions. My premise is that lawyers trying to counteract the animal rights arena must somehow be exposed to the true biological aspects.

Some humane societies and environmental groups sue state and federal wildlife organizations for permitting wildlife species to be harvested or controlled. It seems to me they are confused; usually they should be suing government organizations for not being liberal enough in bag limits or depredation permits to perpetuate, in the long run, even healthier populations with an improved quality of life. Animals, whether pocket gophers or deer, that overpopulate live a pitiful existence and die prematurely from self-limiting factors like starvation and disease. The public needs an empathy for a realistic nature, not just for environments devoid of people. M. V. Hutchins et al. point out that by adhering to a philosophy that emphasizes a reverence for life, you may have to ignore the conditions necessary for life, thus be unfaithful to your own philosophy. Since every environmental issue has trade-offs, too frequently the "humane" donations actually make matters worse.

Animal Rights Vs. Nature

WHEN YOU CONTRIBUTE TO HELP WILDLIFE, BE SURE YOU ARE THINKING ABOUT THE WELFARE OF POPULATIONS OF THE SPECIES AND NOT JUST INDIVIDUAL ANIMALS.

As stated previously, there is a need for environmental and animal welfare organizations, but we also need some way of keeping the extremist animal rights groups from misleading the public about the balance of nature and animal humaneness or we may lose much of the public's confidence about these matters. Animal rights issues provide many opportunities for social conflict because their goals are so extreme. In contrast to animal rights, the growth of animal welfarism during the past few decades has been justified, and animal protection movements have been needed. The wider recognition that it is inappropriate to treat animals inhumanely was long overdue. What concerns me now is whether such animal welfare gains may ultimately be destroyed by extremist animal rights groups.

The exploitation of animals by some people is what opened the door for the antiexploitation of animal rights groups. But extreme animal rightists are about to undo the good they previously accomplished by abusing the confrontational approach. Unfortunately, it is difficult sometimes for the public to sort out legitimate animal welfare concerns from the vested interest ones that require organizations to keep the pot stirred up, since public donations are what provide their livelihood. **MY COMMENT TO THE ANIMAL RIGHTS**

Vested Interests

LEADERS IS THAT IF I COULD AGREE WITH THEIR VIEWS, LIFE WOULD BE MUCH SIMPLER FOR ME, WHEREAS IF THEY AGREED WITH ME, THEY WOULD BE UNEMPLOYED. Of course, it is inherent with advocacy organizations that they literally cannot "afford" to have their propaganda issues resolved, hence the persistent efforts to emotionalize their pet issues and keep them in litigation.

One aspect of the eco-adversary movement is that the leaders must constantly search for new issues to keep their "watchdog" credibility; hence those who make such issues their livelihood are often tempted to blow a questionable issue out of proportion. They assume their rhetoric is unfettered by biological principles, or even common sense. Some groups are very skilled at arousing public emotions, as, for example, in showing a young harp seal being clubbed to death on snow, even though it is very humane to be instantly stunned unconscious and probably the best procedure for harvesting the surplus young of the species. If harp seals happen to be overexploited, then all that is necessary is to reduce the quota harvested each year, but this is not being proposed by the extreme groups. The harp seal issue was a bonanza for Greenpeace after they failed with the nuclear issue.

For various good reasons biologists seldom challenge unsound lawsuits or the eco-guerilla tactics of some environmental extremists. It is no wonder, as J.S. Feierabend points out, why we are forced to accept what public pressure then demands of the state legislatures and Congress, and what the courts dictate. It is almost like an admission of guilt when scientists and professional wildlife managers let these falsifications pass uncontested. Equally articulate rebuttals from scientists are sorely needed, for biologists are much more knowledgeable about these issues than the extreme environmentalists and environmental lawyers.

Since I am being pretty critical of certain types of environmental organizations, let me make it clear that I also think we need honest, dedicated environmental organizations. My derogatory comments apply only to the extreme ones. Too often such organizations have been callous and unconscionable in their propaganda. But the end of the 1980s has shown a change; many formerly extreme environmental organizations are now trying to distance themselves from the irresponsible ones in hopes of regaining the former support and prestige that they once had. I, for one, hope the outrageous activities of the ALF (Animal Liberation Front, founded in 1973 as the Band of Mercy) prove counterproductive. The sad part is that they attract so much media attention without adequate exposure of the other side of the issues. I will let you identify the organizations that fit my charges of using cheap shots and emotional dogma against hunting, trapping, research, agriculture, etc. **USUALLY THE MORE EMOTIONAL THEIR SOLICITATIONS FOR FUNDS, THE LESS LIKELY IT IS THAT YOUR DONATION WILL BE USED TO HELP ANIMALS.**

The manipulation of public desires by creating fictitious balance of nature issues usually does not help the environment.

Vested Interests

In the long run, such action shakes public confidence in trained wildlife specialists, makes the environment the scapegoat, and threatens the future of wildlife conservation by creating controversy between sportsmen and birdwatchers, hunters and anti-hunters, urbanites and agriculturists, yet we all need each other.

The requirements for environmental impact statements (EIS) became law to help protect the fauna and their habitats. How successful has the EIS been? It has been a great help, but does an EIS force a holistic approach? To a degree, discussion of "unwanted" trade-offs usually appears, but too commonly the consultant or other EIS compiler does his or her best to please the client, who naturally is paying for as favorable an EIS as can be contrived, and one that lawyers can successfully use in litigation proceedings.

ENVIRONMENTAL GROUPS PREFER LITIGATION OVER MEDIATION; IT'S HARD TO RAISE FUNDS BY COMPROMISING.

Extremist groups must be against something. Good environmental organizations are needed, of course, to expose environmental insults by individuals, business enterprises, and governments, and they have been very successful with this mission. The problem occurs when they become obstructionists, intentionally polarizing issues, lobbying Congress and agency officials, which most biologists are not permitted to do, and then suing if they don't get their way. Court decisions rarely address the real issues, since it is difficult to develop environmental solutions in court. Also, there are often hidden agendas in the animal rights attack modes.

To raise money, some groups have also carried their approach to civil disobedience. No matter how noble a cause one has, as Margaret Thatcher was quoted in the press, it is not an entitlement to break the law or infringe upon the rights of others. These individuals or groups place themselves above the law and assume that somehow they are absolved from obeying it. It would be self-destructive if such immunity from the legal process existed. A crime is still a crime, even when committed in mass and in good conscience.

An unfortunate aspect of the vested-interest nature of the extremist groups is that it almost makes it mandatory that they be ANTI something, including most local, state, and federal governments. Their approach makes it difficult for them to constructively work with any branch of government to make needed changes in rules and laws. We, the public, are perhaps trapped. We need environmental and animal welfare groups, yet it is our contributions that are progressively making it more difficult for government to lead, even to act judiciously. What is even worse is that some state and wildlife agencies seem most willing to relinquish decision-making to these groups to avoid harassment. Also, individuals from some of these agencies are sometimes "in bed" with the environmental groups. This becomes obvious when key retired state and federal agency employees are quickly hired by

Vested Interests

environmental groups in the same way generals in the military are commonly later employed by defense industries.

Sensationalism is almost inherent in the "anti" approaches employed by extremist groups, whereas the media are seldom interested in reporting common sense statements or corrections to earlier quoted comments by extremists that have been proven false. The situation has become so emotional, and many vested-interest environmental groups are in such dire need of new dramatic issues, that the existing climate makes it difficult for a concerned citizen, or industry for that matter, to reveal a potential hazard of which they may be aware, because it will be blown out of proportion and probably polarized before any constructive steps can be taken. A good example is the almost total ban against DDT. By outlawing DDT, we lost a very effective and the most environmentally safe toxic tracking powder for the control of house mice. Now no company will pay the millions of dollars required to reregister DDT with the Environmental Protection Agency for mouse control because they could never recoup their investment from sales since DDT has passed the 20-year statute of limitations. There are many examples. The public is often confused.

The fraternity of organization decision makers concerning balance of nature aspects of our wildlife heritage is composed of a relatively small number of leaders -- for the most part genuine -- although some are hardly ethical. Decision making, however, is politics, and seldom are the decisions based on the best value judgment possible. The public seldom has a vote; we merely contribute to those environmental organizations that have convinced us with their propaganda that they know what should be done to improve animal welfare, how to avoid further upsetting the balance of nature, and how to save the environment. It appears that even wildlife biologists have become so entrapped in this emotional arena that too often the reality of the balance of nature has been changed into emotional fictitious claims that in the long run hurt rather than help the environment.

WHERE ANIMAL CONTROL PROBLEMS EXIST, THE MOST EFFECTIVE ANSWER IS TO FIND A BETTER ALTERNATIVE SOLUTION.

When faced with vertebrate pest problems, e.g., with depredations from rodents, birds, or predators, just being anti-control does not help. Since so little research has been done to find alternatives to toxic chemicals and lethal methods, it is still necessary to poison rodents, trap coyotes, and shoot deer because such wildlife management and animal control practices can actually humanize nature. When you receive fund-raising solicitations to protect a particular animal species, or to stop the use of guns, traps, or poisons, make certain the plea is legitimate and not just another vested interest "mail scam."

There are many reasons why wildlife managers have difficulty obtaining the public support they deserve. Wildlife

Vested Interests

managers cannot raise money for their cause easily just by doing their daily job. Wildlifers have to think positively -- not negatively; consequently, their objective arguments lose out. They also have to be honest. With the various pressure groups, it is sometimes difficult to know who is managing our wildlife. Some environmental groups seem to think they should dictate management decisions to the professional state and federal wildlife agencies, as appeared to be the case early in 1986 with the California condor. If the wildlife profession doesn't wake up and come to the rescue of Congress and other leaders, extremist groups may be successful in controlling all wildlife legislative decisions. There still are a lot of sensible people out there; biologists, get to work. Unfortunately, the media do not seem interested in asking the right questions of the scientists. For professional biologists to be heard, they must use emotional propaganda, thus risk losing their credibility. I hope this book will stimulate the media to ask the right questions.

EVERY ENVIRONMENTAL SOLUTION HAS TRADE-OFFS.

All environmental decisions should -- in fact, must -- be based on reasonable compromises, but vested interest groups make this difficult to attain. Since the public today demands greater participation in decision-making, it becomes a ripe plum for vested interest groups.

Animal Rights Vs. Nature

The public should analyze the issues before contributing to emotional "anti" animal control appeals. Make sure the arguments for your money are based on sound biological principles. Be suspicious of emotional dogma, which often even appears on the mailing envelope. Unfortunately, the Federal Trade Commission, Postal Service, Environmental Protection Agency, and Internal Revenue Service are not able to closely examine and enforce the frequent tax-free abuses and false claims regarding the effectiveness of various devices organizations may claim will solve wildlife pest problems.

A common fund-raising technique used by many organizations is to dramatize key issues by requesting you to complete a national or international survey, census, or ballot concerning that organization's pet fund-raising issues. The questions are usually so mundane and simple that it is a forgone conclusion how the recipients will answer the questions. But after reading them -- and they are so straightforward that time to reflect about them is not needed -- they usually end up with an appeal for emergency financial support of these critical and vital issues, since by now your emotions should be sufficiently aroused to make you feel benevolent and generous instead of alert.

Since many people claim that the news media provide too much attention to adversary groups, why not have government or a counter-coalition raise pertinent questions that adversary fund-raisers would then be pressured to answer publicly? Doesn't the public have a right to insist that their government expose myths expounded by anti groups? I am

Vested Interests

not aware of any means by which anti groups police themselves to ensure that their claims are substantiated with facts, yet many claim and receive tax-exempt status.

The European rabbit (<u>Oryctolagus</u> <u>cuniculus</u>), like many desert rodents, is adapted to habitats having fairly bare ground and/or turf conditions and does not thrive in lush pastures like alfalfa or clover even if the plants are highly palatable and nutritious. (W.E. Howard)

5. CONSERVATION BIOLOGY

Some new and meaningful "buzz" words are Conservation Biology and Biological Natural Diversity. The goal is to keep natural populations intact and all species of organisms from becoming threatened towards extinction. To achieve this goal, we desperately need a sensible environmental ethic in the national and international conscience that is consistent with the laws of nature. We must learn to live within our environment rather than savaging it.

Without entering the debate over whether or not the "greenhouse effect" is really occurring, we cannot deny that it is wrong for people to pollute the air as is happening today. Our extravagant use of fossil fuel energy, coal, oil and gas, with its carbon dioxide byproduct, is serious. Sure, CO_2 increases the productivity of plants and even reduces the ratio of transpiration of leaves (and water conservation is critical); nonetheless, the potential greenhouse effect (earth warming), and other unknowns are serious. Methane is produced wherever bacteria break down organic matter in the absence of oxygen, and it is said to have 25 times the heat-trapping properties of carbon dioxide, hence our increased production of methane could be a significant greenhouse contributor. Another global heat-trapping pollutant, on a per-molecule-basis perhaps 20,000 times that of carbon dioxide, is chlorofluorocarbon (CFC), which is used as a refrigerant and in air conditioners, the beads in plastic foam, propellants in alcohol sprays, and as solvents for keeping computer chips clean. Our concern with CFCs is global warming and their depletion of the stratospheric ozone layer, which shields the earth from dangerous ultraviolet rays that promote skin cancer, cataracts, weakens immune systems in people and other mammals, and can kill fish and plant life. Nitrous oxide, also a more potent greenhouse gas than carbon dioxide, is also increasing in the atmosphere. It is from automobiles and coal-fired chimneys. Fortunately, carbon dioxide does not change the basic chemistry of the atmosphere as do some polluting gases; it mainly changes the composition.

Most of the world, including the U.S., has been too successful in subduing wild nature. Somehow, we must find ways to develop more self-interest needs for protecting the environment. How can we achieve maximum conservation of biological resources, yet provide a sustainable economic use of the resources? Parks and refuges are too expensive to take care of all these needs. Proper stewardship of the global environment requires balancing environmental conservation with the ever-expanding demands placed upon the earth's natural resources by the growing industrial society and the human population's demand for agricultural resources.

Conservation Biology

Of paramount importance by its potentially dangerous implications is the contrasting of "nature protection" and "nature utilization," for total protection may do more damage to other animals, cause unnecessary suffering of protected species, and even degrade whole ecosystems. Furthermore, most areas contain humans, and such regions are increasing in size. Unless the people can be integrated in the conservation effort, success is very unlikely. Also, the concept of total protection such as in national parks may have the psychological effect which results in the public and politicians thus justifying the exploitation of the remaining areas. While protected areas must remain and be promoted, the exploitation of other areas should be guided by a concept of nature protection through nature utilization. Those who oppose the concept of sustained use of renewable resources basically promote exploitation with its associated long-term problems.

SAVING ENDANGERED SPECIES IS A BUILDING BLOCK OF ENVIRONMENTAL CONSCIOUSNESS.

We do not have a license to exterminate any species, and we should be a servant of this earth. Too often we are victims of greed and self-preservation. It is technically true that society does not "need" any endangered species; however, the environment does not need you or me. Because of the recent economic and social advances made in the current civilization, our newly acquired humanity and ethics demand that we treat all animals better and be concerned about their

Animal Rights Vs. Nature

survival.

Intuitions cannot be used as rational and objective grounds for any theory of environmental ethics. Ethics has its roots in person-to-person relations, and thus human ethics serves as the basis for expanded animal welfare standards which includes nonhuman animals. Only if the norms applied to human ethics give equal weight to every person's value system and at the same time make it possible for each to pursue the realization of his or her own value system in ways compatible with everyone else's similar pursuit, is it possible for such norms to be acceptable to all. A valid system of human ethics, therefore, is a set of moral rules and standards that embody the principle of respect for all persons as persons. Consequently, moral relations have no regard for the particular value system of each person. Animals cannot act morally, cannot be held accountable for what they do, and cannot form judgments about right and wrong. A mountain lion biting the neck and killing a young deer is neither good nor bad, it is just a lion, a functional part of an ecosystem.

People have many views on how to attain supreme enjoyment of nature. Wilderness, for example, means different things to different people. The backpacker dislikes horses; the saddle-sitter disapproves of the 4-wheeler; the cross-country skier frowns on the snowmobiler; the motorist objects to the hazards of cyclists; and mothers protest hunters in adjacent forests. But, please, let's stop making fictitious statements about the balance of nature just to defend our preferences. The only place where environmental "problems" do not exist

Conservation Biology

is in a truly natural area where, by definition, there are no environmental problems since everything that is unaffected by people is considered natural.

RELIGIOUS AND ECONOMIC GOALS OFTEN IGNORE THE LAWS OF NATURE.

It is gratifying that there is a growing concern about nature conservation, but progress would be much greater if ways could be found for conservationists to subscribe to the economists' economic value systems. It is tough to have the world's economy grow yet have our natural species diversity heritage survive. They must not be mutually exclusive. Developing countries cannot risk retarded growth just to protect the environment, so the energy-environmental collision continues. Everyone wants more amenities, even in the developed countries. Unfortunately, from striving to live in a modern world, developing countries are essentially forced to degrade their environment to keep abreast of their growing population and to produce exportable goods to help pay off their enormous financial debts. Food is not grown commercially to fill hungry stomachs, rather it is to make money. Everyone in the world could easily be fed if they had the money to buy the food. The quest for food and fiber is in response to both population and per capita income. The increased demand in developing countries is primarily in response to both population growth and per capita income growth, whereas in developed countries it is mainly just the

Animal Rights Vs. Nature

growth rate.

Economic growth is immediate and blind to higher considerations. Habitat integrity is often in conflict with economic progress. Politics, the urge to be elected, can also be a serious impediment to gaining support for sound animal welfare movements, especially when well-financed animal rights activists are afoot. Since economics rule the world, multiple use of federal land is often governed by economics, not environmental needs.

How can we preserve natural areas if the human population continues to explode? Certainly when it is a matter of human survival, conservation and preservation have little meaning. I have observed many steep rocky hillsides in Mexico that have lost all their soil to erosion just in my lifetime after being cultivated for the production of human food for only a couple of years. In addition to saving our precious soil, I also think it is important that we preserve large areas that are of unusual natural, cultural and historic value, as well as for wildlife, partly because of the indeterministic and hierarchial nature of many environmental systems and the related wildlife issues. With the spotted owl (Strix occidentalis) issue, the real question is just how much old growth timber needs to be set aside. After all, this northern subspecies is NOT an endangered species, only a threatened subspecies, although that is a serious concern.

Elephants in Africa in the past several decades suffered first from loss of habitat, but also due to being overprotected.

Conservation Biology

They exceeded the carrying capacity of many preserves, thus seriously damaging their own environment. Now poachers are the main problem, with payoffs being made to high officials. The current (1988-1990) holocaust and orgy of killing by poachers is tragic. Whether an embargo on all ivory would help, or proper management with harvesting is the answer, is not clear. Ivory cannot be treated economically (supply and demand) like gold, rubies, and other minerals, for it comes from living animals that are becoming scarcer daily, which increases the value to poachers.

It has been a great help to have the World Bank make environmental analysis a central factor when evaluating development loans to Third World nations. The Nature Conservancy in Costa Rica and Conservation International in Bolivia have arranged some very effective "debt-for-nature" swaps, which enable nations in financial debt to be forgiven all or a portion of their foreign debt in return for a commitment to protect needed habitats in tropical forests, grasslands or river basins. The World Wildlife Fund and other organizations are similarly participating elsewhere.

THERE ARE MANY HEROIC EXAMPLES OF CONSERVATION WHERE SPORTSMEN HAVE SAVED WATERFOWL, BIGHORN SHEEP, TULE ELK, WILD TURKEY, ETC.

To protect endangered species, manage wildlife, and

Animal Rights Vs. Nature

control problem animals in ecosystems altered by people, and do so in the most environmentally compatible manner, it is essential that we thoroughly understand how different species of fauna and flora interact in such disturbed environments. We used to say that conservation was the wise use of natural resources for the good of mankind but now conservation is wisely viewed as also being good for all wildlife species as well. Wildlife management is -- or should be -- the science of managing wildlife and their habitat, including people, for the benefit of the entire biota, not just for consumptive uses of game animals. This is why the time has come when others than hunters and trappers must fund such movements. There is a need for a new science of "ecology of modified environments."

People are not too different from animals in the struggle for survival in the balance of nature and, unless economically secure, people rarely spend much energy or funds to help animals for pure ethical considerations. However, most people are willing to make some sacrifice to prevent species from becoming extinct. However, in spite of the unfortunate extinctions of vertebrates during the past few centuries, today many more kinds of animals inhabit all of the continents of the world (other than the North and South Poles) than would have occurred naturally. This increased richness in variety of species is due to the creation of new habitats and intentionally or accidentally introducing many kinds of fish (most of the warm water fish in California are exotics), birds (both song birds and game species), mammals (rats, mice, and a great number of game species), pets and livestock. Some of these species, such as the house mouse, rat, and starling, may not be desirable to some, but that also applies to native species homeowners and farmers call pests.

Since every development or preservation decision will have tradeoffs, social and political issues inevitably emerge. Habitat modification often produces ecological misfits that can be saved only through careful management schemes. Most species are not keystone or critical-link species, and substitution is often possible. New arrivals usually do not cause a net loss of species, but introductions should be made carefully. Small populations that often remain after habitat alterations are vulnerable to genetic drift and inbreeding depression as well as human or climatic changes. To provide harmony and stability in modified environments requires people to assist nature maintain new and acceptable balances.

Nature is not "ending" due to man's activities, as proposed by Bill McKibben, but it is certainly heading for some drastic changes that we surely do not want to happen. He is correct that man's influence on nature is leading to the imposition of an artificial world in place of the broken natural one. I also agree that we must limit our numbers, our desires, and our ambitions so nature can someday resume its independent working.

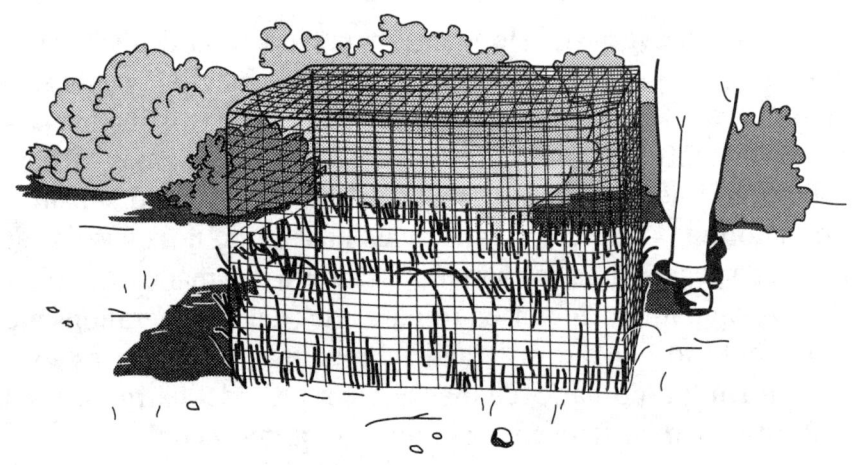

Following a broadcast seeding of grasses and legumes in a stand of chemise (<u>Adenostoma</u> <u>fasciculatum</u>), the only seeds that germinated were those protected from small rodents by a cage. Once such areas are revegetated, the same species of rodents decline with the change in the habitat even though palatable food becomes much more abundant. (Pasture and Range Research Techniques, Comstock Publishing Associates, New York, 1962. Photo by W.E. Howard)

6. WILDLIFE MANAGEMENT ANIMAL CONTROL

Wildlife is a resource belonging to the public, and the need for more environmental sensitivity today is obvious, due in part to the animal rights movement. People have always exploited animals and had conflict with wildlife. This issue is currently in another dramatic period of transition. **IN ENVIRONMENTS MAN HAS MODIFIED, NATURE SELDOM "KNOWS BEST."** In disturbed environments wildlife must be managed or controlled to protect other species or resources, for public health reasons, or because someone currently views them as a pest, whether or not they are actually doing damage, such as a skunk or rattlesnake under the back porch.

Most vertebrate pests are pests because they are so well adapted to a wide range of modified habitats. Also, they have high resilience, and their populations promptly recover after control operations are reduced. They are pests because they are difficult to keep well below the carrying capacity, but an animal that may be a serious pest to one person may at the same time have aesthetic value to someone else.

Animal Rights Vs. Nature

Few people realize that agricultural crops and most home landscaping would not survive if native mammals were given free range. Agriculture cannot succeed with free-ranging native mammals. This is because these plants and the animal pests did not co-evolve and develop resistance to each other and the crops are dispersed as part of a new plant community. "Pest" is a good term for a problem animal, as R.M. Timm points out, because it is a subjective term that permits another person to have a different opinion.

Vertebrates become pests when they compete with people or otherwise become an annoyance. This occurs because man has encroached upon their environment to provide himself with food, fiber, other resources, a desired lifestyle, or to avoid public health problems. Modifying habitats produces ecological misfits, e.g., endangered species that can be saved only through management schemes. The public needs to recognize that pests are a necessary segment of everyday life, and all organisms have other organisms that are pests to them. The public has a right to demand from animal control experts the highest standards of wildlife management and conservation, and that they react in a humane and responsible manner. Animal damage control is an integral part of life, and it helps establish true harmony between man and nature. Government animal damage control personnel are frequently called in to help save endangered species such as the whooping crane, desert tortoise, San Joaquin kit fox, California least tern, peregrine falcon, etc. Proper animal damage control establishes morality in "the role of man in nature," as will be brought out in this book. A zealous

Wildlife Management/Animal Control

"protectionist ethic" is unsound biologically, and conservationists need an animal damage control and wildlife management ethic based on the laws of nature, not emotionalism.

There is a tendency by many conservationists to erroneously consider everything "chemical" as unnatural and bad, while everything "biological" is intrinsically desirable. Both are false. What is natural is exciting but not necessarily the best as far as most people are concerned, especially if they have been injured by an earthquake, volcanic eruption, hurricane, suffered an economic loss from wildlife, or have ants in their kitchen.

We are all hypocrites. It seems that most of us want what was here before Europeans arrived in North America, **EXCEPT** for those things that affect us. We don't want to landscape our homesteads with what grows naturally; they are just weeds. We don't want most native animals in our homes and gardens; they are pests. Yet, too often we say that farmers, livestock operators and foresters should encourage all kinds of wildlife. Some people like to think that wild animals have as much right to live where they want to, except where the animals would be a pest to them. They forget that in nature many animals are possessive and prevent other individuals from occupying their habitat, the same as do people. That is nature.

Not only do some surplus populations of wildlife compete with us for food, destroy the products of our labor, cause zoonoses or become a nuisance, but they may also seriously impact the environment before dying nature's way. Even though rats usually do not die from diseases, the potential of the deadly human disease of pneumonic plague is a biological time bomb we are facing in the greater Los Angeles area where plague is endemic and because susceptible ground squirrels live close to people. When ground squirrels get plague and die, their infected fleas may be picked up by dogs, cats, and rats, then carried to humans. With the incidence of pneumonic plague annually becoming more frequent, this airborne form of the Black Death of Europe, which kills people in 1 to 3 days, could spread like wildfire in parts of Southern California and other areas of the United States because of the mobility of people.

Life is not without risk, but to determine a proper risk-to-benefit ratio is difficult. Few people stop to think what they consider an acceptable risk in the food they eat, the medicines they take, and the lifestyle they choose. It is only natural that we are always more concerned about involuntary risks than those we take voluntarily. Sports, driving cars, using gas and electricity, and public transportation are risks we usually leave to fate. Also, we must remember that all environmental solutions have trade-offs.

Naturally evolved animal-plant communities are often not delicately balanced, except that predators depend upon prey species for their subsistence. Natural prey populations are less dependent upon predators, but when preyed upon they are then usually maintained at a healthier density. Evolution has seen to it that natural predation helps maintain an optimum

Wildlife Management/Animal Control

density of their native prey species. The removal of individuals and often even populations of specific wildlife species, when they become serious pests, usually does not pose serious adverse ecological problems for other animals in the community, except for dependent predators. The removal of a population such as deer over a large area might pose a serious problem for wolves, as the removal of prairie dogs (<u>Cynomys</u> spp.) would affect black-tailed ferrets (<u>Mustela</u> <u>nigripes</u>). Similarly, the removal of herbivores in time may permit marked changes to occur in the vegetation.

Since we are a part of the balance of nature, what is our role? Should we be willing to share our agricultural crops with wild animals, insects, and weeds, thus requiring that more land be devoted to food production? If we shared our agricultural crops with wildlife, we would probably discover that there is not sufficient arable land available for us to feed ourselves. The surpluses would soon disappear. Consumers in the United States now have access to food on the most favorable terms ever. For agriculture to retain its domestic markets and expand its foreign exports depends on its ability to lower production costs and hence share less and less of the crops to pests. It is economics. Some developing countries occasionally lose more than half of their crops to wildlife pests.

Toxic chemicals and other lethal control methods for managing problem animals are rightfully coming under increasing fire by animal rights activists and environmentalists. Unfortunately, the needed nonlethal, nontoxic chemical

Animal Rights Vs. Nature

alternative control methods, that even farmers and ranchers prefer, are not available due to inadequate research funds. Everyone recognizes the need for developing strategies and methodologies that minimize suffering and environmental contamination. What bothers me is that none of the organizations which raise thousands of dollars to ban traps and other control tools ever offer to help fund research for more desirable control methods. I guess this is to be expected, for how many of us would look favorably on research that might cut off our bread and butter?

By definition, without people there can be no animal or nature problems, for then everything is considered natural. However, we must work with people, and I find that people's perceptions about what is right or wrong concerning environmental issues usually depend upon the manner in which the animals affect them. Further, we live in an economic world, and everyone -- including those who govern our animal resources -- is caught up in an economic, sociological, and political quagmire concerning animal issues. When it is a matter of human survival or one's own affairs, wildlife preservation and conservation have little meaning, and we must remember that ultimately the consumer pays for all animal damage to crops and other resources. Therefore, we need to learn how to economically manage and control wildlife in socially acceptable ways in the environments people have modified. One difficulty is that it is hard to show economic benefits for many good environmental activities.

There is a need to appraise the changing attitudes that

Wildlife Management/Animal Control

affect animal control, both beneficially and detrimentally, and to analyze what action can be taken to constructively modify some of the negative attitudes about animal control. If we are going to save our wildlife heritage in a growing human population, where most environments have been altered by people, we must assume the responsibility of managing and controlling wildlife rather than leaving their fate to the whims of nature. Only by managing and controlling wildlife can we establish true harmony between people and the fauna. If one opposes any use of wild animals, I wonder **HOW THEY PROPOSE THAT THE SURPLUS ANIMALS SHOULD DIE** in human-disturbed environments.

Once a habitat has been changed by man, to ignore necessary control measures to keep certain vertebrate species in balance is to invite ecological disharmony of land. To try to protect all vertebrate "pests" in the interest of conservation may actually be working against the very goals striven for. Many logged-over coniferous forests in the West have reverted to a brush field because either the conifer seeds used to establish the next stand were not treated to repel deer mice, or these seed-eating forest rodents were not kept at their former low levels with poisons until the new stand of seedlings was established. Now, following logging or wild fires seedlings are planted. To ensure successful regeneration it may still be necessary to protect seedlings from deer, rabbits, porcupines, pocket gophers, and other rodents.

When we protect too many feral burros in the West, the native mountain sheep may become the scapegoats when the

Animal Rights Vs. Nature

burros deny the mountain sheep access to their former water holes.

If wildlife needs to be managed and controlled in an altered environment, it is a legitimate question to ask: "How much should we manage?" Sometimes in wilderness areas and remote regions in national parks the best management is to do nothing. Leaving it to nature is fine when it has first been determined that the best management scheme for that situation is to let nature's survival-of-the-fittest establish a new balance. However, when our objective is to improve the welfare of individual animals or their populations in modified environments, we should be prepared to also manage wildlife numbers.

ONCE MAN HAS MODIFIED AN ENVIRONMENT, HE HAS A MORAL OBLIGATION TO ACTIVELY BOTH MANAGE AND CONTROL THE WILDLIFE SPECIES PRESENT

If you analyze wildlife problems, you will note that there are no clear-cut, right-or-wrong environmental answers of how to best manage any wildlife issue because the "correctness" of almost all decisions depends upon one's perspective. Biologists often point out how excess animals are destined to die prematurely, but they seldom stress the absolute necessity that there must be many deaths in order to keep any population healthy, nor do they elucidate how man might promote

Wildlife Management/Animal Control

the fairly high, yet beneficial, required death rates.

Carrying capacity is not just the number of animals an area can support, but more correctly it is the not easily defined ecological optimum density for each species the habitat can support and perpetuate throughout each year without damage to the animals nor the habitat. When wildlife exceed the carrying capacity, as they frequently do, the excess animals die. However, the greatest impact on wildlife comes from urbanization, roads, swimming, boating, water-skiing, camping, hiking, picnicking, and other social activities; whereas hunting, trapping, and fishing produce less negative impact on wildlife.

When people overprotect wild animals and the surplus populations develop diseases or cause serious aggression against their own kind, it is similar to what would happen if humans contracted highly infectious and contagious diseases and were not isolated but allowed to mix with other people.

Most farmers like the out-of-doors and are fond of wildlife as long as animals are not destroying too much of their crop. During 43 years with the University, I have had many requests from agriculturists as to how they might leave woodlots, ditch-bank vegetation and other habitats for wildlife without producing economic pest problems. In many instances this would be feasible if county agricultural agents were able to monitor these sites each spring to determine the breeding potential of the rodent populations that survived the winter. If the agent found that so many rodents survived the winter

Animal Rights Vs. Nature

that they posed an economic threat to the coming summer crop, then the agricultural agent should be encouraged to apply enough rodent bait to reduce the density of the threatening pest populations. Such action would require very little poison bait because the areas to be treated are usually small, and these scattered populations of rodents merely have to be reduced enough so they do not have the reproductive potential to cause damage the next summer. But since many environmental groups insist that growers not poison until after a serious pest problem has developed, growers are usually forced to wait to conduct control until a high rodent population has developed in a crop. This often means that hundreds, even thousands, of times as much poison must be used as would have been needed in preventive control. Also, once rodents have started feeding on a crop, they are more difficult to control. Unfortunately, many environmental groups cannot condone the use of poisons, even when biologically desirable, as they often need an "anti poison" image to raise money. At times these groups are forced to let wildlife die and the environment suffer since they must appear to be anti all killing.

Many people become disturbed about the possible extermination of local animal populations, as if such local eradications would have devastating consequences on nature's balance. This is unfortunate because in intense agriculture and urban situations the complete local elimination of certain serious pests is often the most desirable goal from both an environmental and balance-of-nature standpoint. Homeowners tolerate very few wild species. The large game animals and

Wildlife Management/Animal Control

large carnivores have usually been eliminated from intensively farmed areas.

I find it both ridiculous and environmentally disrupting, for example, to allow species of field rodents in intensive agricultural areas to build up to serious economic densities before they can be controlled. It is then necessary to use much larger amounts of rodenticides, and this leaves the predator populations that may have built up with little available prey. Such artificial cycling of predator and prey populations is both inhumane and environmentally unsound. In these situations, pest species should be locally eradicated "when feasible," or held at very low numbers because control operations would then primarily involve just monitoring new arrivals. With rodent pests, their presence can sometimes be successfully monitored using nontoxic bait. Toxic baiting would then be required only on a very limited basis to control invading individuals since populations would not have had time to develop. This is called permanent preventive control.

AGRICULTURAL CROPS AND HOME LANDSCAPING CANNOT COEXIST WITH FREE-RANGING NATIVE MAMMALS.

Most agricultural and landscape plants are exotics, and cohabitation with native mammals has not had a chance to evolve. One exception may be kelp. It does appear that if we had all the seals and sea otters that used to be present on the West Coast, there would probably be fewer sea urchins, hence more kelp to be harvested. Most wild grazing and browsing herbivores are kept out of agricultural crops and people's gardens for obvious reasons. Thus wildlife damage control is an integral part of life, for it can help establish a better harmony between man and nature in modified environments; but control practices should follow life-sparing approaches whenever feasible.

Integrated pest management (IPM) is practiced in the control of wildlife pests, for it is often cheaper and simpler to use nonlethal control methods whenever possible. The animal control methods usually tried include such nonlethal approaches as sound, visual frightening devices, encouraging biological control with predators, habitat manipulations to discourage pests, repellents, construction of barriers, etc. The lethal methods, e.g., shooting, trapping, and poisoning, involve many more regulations and restrictions, which fortunately discourage their use unless absolutely necessary. Potential biological control methods of controlling animal damage include: (1) encouraging predators of the pest, which usually is not a solution but may help; (2) building them out or making the habitat unsuitable for the pest to survive; (3) introducing or augmenting diseases (epizootics) against them; (4) using chemosterilants to keep pest populations in check; and (5) introducing genetic traits that will make the populations of the pest species less successful. Ultrasonic devices mounted on automobiles or used in homes and warehouses have not been successful.

Wildlife Management/Animal Control

One aspect of agricultural monoculture that is usually not recognized is that there are far fewer <u>vertebrate</u> pest species in monoculture than in diversified agriculture, even though biological diversification can result in greater stabilization of nonagricultural environments. Many ecologists complain that monocultures break down the ecological systems and may produce disastrous populations of insects and other pests, such as the potato blight in Ireland. This has to be qualified when referring to wild vertebrate species. There are very few vertebrate species that can live in monocultural agriculture. Many can feed in monocultures seasonally or for short periods of their daily routine, but few can live exclusively in monocultural fields. In the Sacramento and San Joaquin Valleys of California, for example, if alfalfa is planted over very large areas, no native mammals can thrive there except voles and pocket gophers. Dozens of other rodent species and other native animals that used to live there are excluded. However, either the voles or pocket gophers could become more abundant per-unit-area in alfalfa fields than all the other species of rodents and other mammals that may have occupied that same terrain under natural conditions.

Self-regulatory feedback mechanisms in nature tend to ensure the sustenance, adaptiveness, and perpetuation of a <u>greater</u> variety of vertebrates in varied habitats, thus making their control more difficult in diversified agriculture. Hedge rows and riparian vegetation along ditches can create many agricultural problems with rodents by providing suitable habitat between crops when the plowed fields do not provide food or cover for most animals. The conversion of natural habitats to agricultural crops eliminates many species of vertebrates locally, although a few species may profit by the new habitats and become unusually abundant. However, their artificial control will still be much simpler than trying to control many more kinds of animals that might be present if the fields were composed of diversified agriculture.

It may come as a surprise to learn that habitat modification should not be employed a priori to control wildlife pests. It seems like a nice humane way to address pest problems, but artificially altering habitats so a pest can no longer live there is almost always an undesirable form of biological control to use against vertebrates. It is usually more ecologically advisable to use a toxicant, for any habitat alteration may be more disruptive to nontarget wildlife than even callous use of lethal control methods. Whenever a habitat is modified, most of the species of vertebrates living there tend to suffer. The suitability of habitats is what determines how well a species thrives, and any modification will usually make the habitat less suitable for most native vertebrates that formerly occupied the area.

Because of evolved natural self-limiting forces, the density of wild vertebrate species in favorable natural habitats cannot be artificially increased very much by providing additional food or cover or by reducing predators. Some herbivores will be an exception and temporarily attain increased densities when fed artificially.

Many animals are favored by habitats with sparse

Wildlife Management/Animal Control

vegetation. This is a strange phenomenon. Species such as the European rabbit, California ground squirrel and most desert rodents do not thrive in lush, herbaceous vegetation. Where weather and soil conditions do not favor much herbaceous vegetation, many native vertebrates and invertebrates can then maintain bare-ground conditions where they thrive best. If there is a good response to range fertilization and/or irrigation, many rodents and European rabbits that formerly lived there may then no longer thrive. Unfortunately, this approach cannot be used to control pests because of its cost.

The objective of animal control is to reduce a problem, such as depredation to a crop or other resource, or a pest to a homeowner, whether the method be with frightening devices, repellents, chemosterilants, traps, guns, or toxicants. At times the goal of reduction may be zero individuals, as with rats in a house, moles in a lawn, pocket gophers in a citrus orchard, or burrowing rodents in narrow flood-control levees. When the need for reduction is indicated, the level of density considered tolerable is that density which is fully consistent with the factors that raised the particular vertebrate species to a pest status in that situation. Whereas wildlife management has largely been based on "use syndromes," wildlife control is more a consequence of health and economic survival, or for the protection of endangered species. An objective of management is to ensure that the species survives in adequate numbers to play its role in maintaining the health and stability of the ecosystem, and that harvest may occur where consistent with the above primary objective. Management is complicated, with a need to understand and estimate the carrying capacities, whereas control is usually to prevent damage or the reduction of a local population to a tolerable level, as determined by the welfare of the factors that control has been undertaken to protect.

When vertebrates are <u>managed</u>, the objective favors the well-being of local populations of the species in question, whereas a vertebrate <u>control</u> operation has primary benefit factors other than the individual or species being controlled. Rat control, in a home or elsewhere, is not concerned with the welfare of rats. The main objective of deer control in a forest plantation is to protect new trees, although the control procedure adopted may incorporate deer management considerations. When there are too many deer in a new forest plantation, wildlife managers may want to manage the deer for the welfare of the deer, whereas at the same time the forester may want the deer controlled to protect the forest. In these situations, the same action -- increased hunting -- can be taken to satisfy both management and control needs.

Where large species of prey and predators are involved, the public seldom understands that it is often desirable to manage both. Examples include the wolf-moose populations in Alaska and the mountain lion-coyote-deer populations in California. Man needs to become a predator of both prey and predator in these instances to maintain a balance in the population densities.

Predator control -- in fact, all animal damage control as now practiced in the United States -- has no significant effect

Wildlife Management/Animal Control

on the basic flow of materials and energy in the environment or threatens the survival of a species. Modern vertebrate pest control operations do not damage the welfare of wildlife communities in man-modified environments; in fact, the implementation of a mortality factor under regulations often helps create a more favorable balance, and many predators are pretty nasty compared to human predation.

The fact that most carnivores are at the top of the food chain means they are <u>not</u> as important ecologically in the food web as primary producers and consumers. Also, hunters and trappers, in a biological sense, can be effective and relatively humane predators, thus compensating for the loss of large carnivores. Most carnivores actually contribute far more carrion to an ecosystem from their uneaten kills than they consume from finding dead carcasses. Even though most carnivores appear to be extremely cunning predators, it is amazing how <u>inefficient</u> they are in reducing their native prey (except for domestic livestock) to very low levels, except in some altered environments. Of course, to do otherwise would be self-destructive. Those that were highly successful in capturing their major prey have become extinct as a result of destroying their food supply.

Due to man's modification of the environment, it often becomes a necessity to control some predators to protect endangered species and improve the quality of life for many other species. We have come a long way from where "a good coyote or wolf is a dead one." But, as before, extremism still flourishes on both sides and inadequate attention is given to

Animal Rights Vs. Nature

the laws of nature. The public is confused. The first European settlers of the United States quickly learned that economic ranching was impossible without wolf control and that all forms of agricultural production required protection from many species of wildlife.

It is very inhumane to encourage the re-establishment of wolves, grizzly bears, and other predators where they are likely to feed on domestic animals that have had their natural escape mechanism bred out of them. It is further inhumane because these depredators then have to be removed. There have been areas in the greater Los Angeles area where if you allow your dog or cat to run free, even briefly, it may be taken for a meal by a coyote or other predator. Even though human predation operating under regulations can be considerably more humane than natural predation, we must still do all we can to minimize pain and suffering of our prey victims.

Healthy reproducing populations of bears, mountain lions (<u>Felis</u> <u>concolor</u>), and wolves cannot tolerate human communities any more than domestic livestock can cohabit with high numbers of these carnivores. There is little sense in reintroducing or encouraging recolonization of grizzly bears or other large carnivores into their former range unless there is a good chance that they will reproduce, which means it is inevitable that they will multiply beyond the carrying capacity of the habitat with the surplus dispersing.

Mountain lions (puma, cougar) are highly secretive top carnivores that cannot live in close proximity to people. All top carnivores have evolved a high degree of territorialism to prevent the species from destroying their food supply. This means the subdominant animals must disperse and usually die rather than remain and face an intraspecific battle. Today when surplus lions in California try to disperse, people force them back, hence the overpopulation of lions with suspected high rates of cannibalism, overutilization of some deer herds, malnutrition, and intense fighting. When the natural mortality mechanism for lions, i.e., dispersal, is lost, the whole population of lions suffers miserably just like unchecked populations of livestock, deer, or other animals.

An overpopulation of bears, in fact of any large carnivore, can be dangerous if humans are also numerous. Since 1975, 13 children have been attacked, 1 fatally, by coyotes in Los Angeles, California. There has been no hunting of mountain lions in California from 1971 to 1990, and in 1987 a 5-year-old girl and a 6-year-old boy were seriously mauled by lions. In 1989 a 5-year-old boy was killed near his home by a mountain lion in Montana. They are dangerous predators when territoriality forces individuals to disperse and find food elsewhere.

Much more research will be needed before the use of natural predators will become a useful practice for abating animal damage since before most pest species are no longer a problem, they must be reduced to a much lower density than what predators can achieve. Predators are not species-specific in their prey selection; most are opportunists.

Animal Rights Vs. Nature

Even though it is doubtful that the elimination of coyotes would have any undesirable ecological consequences, no one proposes their extermination. The objective of those who suffer livestock losses from coyotes is to stop the depredations of fairly helpless domestic animals. Where guarding dogs or fences work, they are often used. But in many areas, traps, snares, toxic baits, and shooting are still required to control coyotes. Sometimes the eradication of all of a small number of local coyotes is the best solution. In our experiments, when we removed the usual killer coyote, another of the small group of coyotes involved immediately became the killer. Sheep killing increases when coyotes have pups.

Since people can largely dominate all wild species and have some control over nature, what is our moral/ethical responsibility with wildlife populations that have been adversely affected by our actions? Should we let them be governed as much as possible by the existing natural forces, even if it inflicts suffering to the animals, or should we manage the population as best we can so as to keep them in as healthy a reproducing state as possible? Shouldn't we use our dominance to manage and control populations of wildlife for their welfare?

Every environmental solution has trade-offs that must be considered; there are no moral or ethical rights or wrongs about environmental issues. However, we must work with people, and **PEOPLE'S PERCEPTIONS ABOUT WHAT IS RIGHT OR WRONG CONCERNING ENVIRONMENTAL ISSUES DEPEND UPON THE MANNER IN**

Wildlife Management/Animal Control

WHICH THE ANIMALS AFFECT THEM. The so-called superiority of natural processes over man's management schemes as claimed by some depends entirely upon one's viewpoint and the current circumstances. In modified environments, management by man is usually superior to natural processes.

Repeated stocking (artificially releasing individuals of a native or exotic species) is usually unnecessary in good wildlife habitat except to initially re-establish a population or, as is common practice, to seasonally increase the supply of fish and game available to sportsmen beyond the normal carrying capacity of that environment for that species. Sport hunting is used as a management tool in the U.S. to remove some of the excess animals of a population without damaging the breeding stock. It is a good wildlife husbandry practice and does a better job than the few remaining natural predators, yet does not exclude these predators. Wildlife management programs for game and nongame wildlife in the U.S. are financed mostly by sportsmen's license fees and excise taxes on the sporting equipment they use. Sport hunting in recent decades has not led to the extinction of a wild species, nor caused any to become rare or endangered; instead, hunting provides money for their protection, habitat management, and helps regulate their population densities.

When managing wildlife the question arises, "Should people _use_ animals for food, fiber, and pleasure?" Isn't it the "natural" thing to do? All life utilizes other life. The difference between human use of other kinds of life and how wild

Animal Rights Vs. Nature

animals use other animals is that man possesses a conscience and has moral concerns. Consequently, man should not inflict useless, unnecessary pain on animals any more than he does on other humans. It is sometimes necessary to use animals in various experiments, but every effort must be made to find alternatives to such experiments.

If the "balancing" forces of nature were better understood by the public, both consumptive and nonconsumptive uses of wildlife might be optimized for the benefit of both people and wildlife populations. The public needs an empathy for realistic nature, not just an idealistic affinity for environmental conditions without people. As stated by S.R. Kellert, the common ground will be the fundamental search for an ethic of the land and its living components that embraces both science and human considerations.

Overprotection in the management of wildlife species, even in national parks, which are really only biological islands, sometimes works against the very goals being striven for, as all species of wildlife are predisposed to overpopulate and then become self-limiting by stressful means (starvation, disease) that we usually consider undesirable. There are places where illegal poaching of African elephants is still serious; but in other herds, the problem is clearly one of too many elephants for the amount of land available to them. Where the overabundance of elephants in some African national parks results in a sharp loss of trees that they need as food, do you think the surplus elephants should be shot for the welfare of the herds and be utilized by native people?

Wildlife Management/Animal Control

In 1972 India passed a wildlife protection act, making it illegal to kill any animal. This is "overprotection." Tigers and elephants, like all animals, do not recognize man's property lines. As their populations increase in national parks and forest reserves, more and more tigers and elephants are forced by their own kind to forage outside the parks. Consequently, many villagers are killed by tigers and even elephants. In 1982 when I studied this problem, 60 Indian wardens of tiger preserves and national parks calculated for me that about 150 villagers had been killed by surplus tigers just in the past year.

One of the wardens I worked with had a badly scarred face. After we got well acquainted at our 3-week training program in Kanha National Park, Madhya Pradesh Province, he asked me to come out into the forest and he would explain what happened. I assumed he had been attacked by a tiger. However, he told me that he had driven outside of the park with three of his staff to look at a boundary problem. Upon their return to the national park, they stopped in a village for some items they all wanted, for they rarely went to the village. When the villagers realized that none of the four was armed, they attempted to kill them. When the police arrived, the warden's heart was still beating and his life was saved. But the three staff members were not so lucky. It is tragic when people surrounding the national parks hate the park so much that the wardens and other officials are unable to venture outside their boundaries without armed guards or carrying weapons themselves.

Animal Rights Vs. Nature

OVERPROTECTION IN THE U.S. OF WOLVES, GRIZZLY BEARS, AND MOUNTAIN LIONS CAN LEAD TO THE SAME PROBLEM INDIA HAS FROM OVERPROTECTING TIGERS AND ELEPHANTS.

Foolish management of wolves, grizzly bears, or mountain lions could create similar resentment in the U.S. Environmentalists must recognize that large populations of people cannot successfully cohabit with some of our carnivores any more than with tigers and elephants. Thus, once wolves or bears -- if locally classified as an endangered or threatened species -- leave their geographic sanctuary, they should then no longer be considered endangered but automatically regulated by that state's game and animal depredation laws. Such animals must be removed; they cannot be successfully translocated.

How important is it that we not exterminate any species? From a moral and ethical point of view, I certainly want to preserve a viable population of all fauna and flora. But how much human life-threatening emergencies are necessary before the saving of any wild species is unjustifiable, if any? **OF THE HUNDREDS OF THOUSANDS OF SPECIES THAT HAVE DISAPPEARED FROM THIS EARTH, HOW HAS THE LOSS OF ANY OF THEM HAMPERED CIVILIZATION?** On the other hand, it seems fairly clear that modern civilization could not coexist with saber-toothed tigers and mammoths, unless confined in parks

Wildlife Management/Animal Control

or wilderness areas, not any more than the former freely migrating herds of American bison could be tolerated today. In the interior valleys of California, it would not be wise to attempt re-establishment of tule elk, pronghorn antelope, mule deer, grizzly bears, and beavers in their former haunts. There are too many people present.

Even if the loss of a species is not like removing the pillar from under a bridge, and although the disappearance of certain species might seem like "good riddance" to many, **MOST PEOPLE DO NOT WANT TO SEE THE COMPLETE LOSS OF ANY BIOLOGICAL INFORMATION THAT IS INHERENT TO EACH SPECIES.** However, the recent manifestation of social attitudes towards overprotection of wildlife must also be dispelled for the welfare of these same wildlife species.

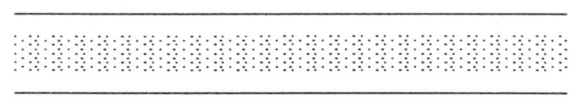

The following questions will help you test your ability to be objective about managing and controlling wildlife:

1. Since we can largely dominate all species and have some control over nature, what do you think is our moral/ethical responsibility to wildlife populations that have been adversely affected by our actions? Assum-

Animal Rights Vs. Nature

ing the people are going to remain, should we let the animals be governed as much as possible by the current natural forces, even if it inflicts great suffering and even local elimination of some species, or should we manage the populations so as to keep them as much as possible in a healthy balanced state with quality of life?

2. In modified environments, can man respond to wildlife's needs more rationally and ethically than nature since nature seldom emulates good wildlife husbandry practices and does not operate under humane regulations?

3. Once people have disrupted the natural balance, do we have a moral obligation to correct the problem as best we can by using the most humane management procedures available?

4. Do people have a moral obligation to protect from cruel predation animals they have domesticated?

5. If you think it wrong to prevent coyotes from killing lambs, cats, and dogs for humane reasons, do you think people should stop raising sheep or having pets in such localities instead of trying to control the coyotes? How do you suggest children should be protected from coyotes?

6. Is it better to control a species that becomes so exces-

Wildlife Management/Animal Control

sively abundant in a human-modified environment that it threatens other species, or should we let nature take its course, even though we know it is impossible to re-establish the original habitat and fauna?

7. What liberties will you grant those who must manage and control competing or predatory species when such action is needed to save an endangered species?

Regenerating conifers can be killed or held back 10 to 20 years when overbrowsed each year by high populations of deer. Such trees are safe only after the terminal escapes and gets out of reach of deer. Kaibab National Forest, Arizona. (W.E. Howard)

7. HUNTING AND TRAPPING

The oldest and original phase of all known human cultures was the era of hunting and gathering. Most of the historic existence of humans was spent as hunters and gatherers. Only in the last phase has agriculture and husbandry been developed. Hunting is thus the oldest profession of humans. In several highly civilized countries hunting has developed to a complex hunting culture. Yet in just such countries we find especially rich wildlife populations today because the species were wanted by the hunters. Trophies were common even in the Stone Age. No one knows how many species of animals were exterminated by early man's hunting for food and self-protection. In modern times, extermination has been mainly due to habitat destruction. In historical times we have the evidence of the Maoris of New Zealand eliminating all the genera of the flightless moa, the largest of which stood about 3 meters. Hunting regulations in the U.S. came about because "unrestricted" hunting -- especially market hunting -- clearly threatened many species of mammals and birds. Today such overharvesting is no longer a problem because, should it occur, extra protection for a year or two by state game or

Animal Rights Vs. Nature

conservation organizations, financed by sportsmen, permits such populations to quickly recover. Illegal harvesting of wildlife is still a minor problem; but because of the license fees that hunters and trappers pay for the privilege of hunting and trapping, wardens are hired to enforce the various state bag limits.

Fortunately, overall there has been a fairly happy marriage of the hunting fraternity and conservation-management agencies that has helped ensure the survival of game species and their habitats. Because of other pressures, state fish and game conservation agencies are actually quite conservative in their game management and protection regulations, sometimes too much so for the best health of the species being protected. It is important for the public to realize that it is primarily the hunters' monies that make this conservation effort possible. If these funds are eliminated by organizations opposed to hunting, no one seems to know where other sources of revenue to support the conservation agencies would come from. Certainly those opposed to hunting are not offering to contribute to finance the needed wardens, although hopefully this will soon change.

Let's look at these questions: Is hunting brutal, sadistic, and primitive? Or is it an effective and relatively humane way of helping nature maintain healthy animal populations in modified environments?

A simplistic, emotional way to look at hunting is to ask how anyone can morally justify and enjoy killing wild birds

Hunting and Trapping

or mammals. First of all, "**Hunting**" is not equivalent to "**Hunters.**" The motives of hunters vary widely but none explains what hunting is anymore than the motives of those who become doctors describe medical sciences, or the motives of those who become lawyers describe the judicial systems (Hennig 1988). The human hunter and the animal predator have much the same interaction with the prey, except people predate under regulations, not nature's no-holds-barred, survival-of-the-fittest predation. When done with the proper respect, nature surely accepts regulated hunting and trapping as appropriate actions even though it causes some pain and suffering.

Hunters find great satisfaction in the roles they play with nature; i.e., preserving the basic population, managing a sustained form of production, and utilizing the natural recruitment. Hunters harvest selectively and ensure a sustained yield. Their laws regulate the seasons, set quotas, close seasons when necessary, and regulate the methodology used in taking game and the care and disposal of animal by-products. Unfortunately, it is not feasible in the long run to think that sufficient fish and game can be provided for all in pursuit of this sport. Many return empty handed but enjoyed being outdoors.

On the other hand, upon careful analysis of nature in altered environments, a good case can be made showing how nature's balancing process demands both a high premature mortality rate and meat eaters. The balance of nature and nature's food web is based upon everything feeding on other species and in turn being consumed. Pain and suffering are an integral part of nature, demanded by evolution and the survival-of-the-fittest process. However, even though we are part of nature, this still does not give us a license to inflict "unnecessary" suffering; i.e., we should make the lethal management tools as humane as we can, yet still play the essential role of a predator assisting nature in modified environments by harvesting the surplus that otherwise would damage the environment and the species' own welfare.

HUNTERS AND TRAPPERS CAN HUMANIZE NATURE.

It is interesting how most people condone the natural brutality of nature's predators cruelly feeding upon fairly helpless prey but object to hunters humanizing these events, even when the need for such predation is the consequence of our modifying the environment, thus increasing the exposure of the prey species.

Nature continues in her survival-of-the-fittest manner no matter what various species, including man, may have done to other species. When one species leads to the extinction of another, Mother Earth loses diversity. And, unfortunately, the evolution of new species is extremely slow and the rate of extinction is accelerating at a frightful pace. Modern hunting is not responsible for extinction; rather it provides the funds to buy habitats, hire wardens, and employ biologists to ensure that healthy populations in suitable modified environments still

Hunting and Trapping

persist.

Hunters today provide a sustainable utilization of wildlife populations through conservative hunting. To produce a utilizable population increase, it is paramount to protect or conserve the basic breeding stock (population) to produce further utilization. Therefore, utilization and preservation are inseparably entwined. Depending on many external factors, the ratio of the level of utilization and the basic population may change from year to year. No one can deny that hunting under regulations is the humane way to control most game species that would otherwise overpopulate and die from horrible species self-limiting starvation, disease, etc.

It is not easy to analyze the humaneness of wildlife management, or the harvest or population regulation of these animals through hunting and trapping. It seems obvious that opposition is primarily from vested interests, personal philosophy or emotion rather than based on biological facts, for in modified environments a desirable harmony between people and the fauna can be established only if animal populations are managed. In disturbed environments, where the original predator-prey balance can no longer exist, nature needs people as a "predator" to prevent excessive species self-limitation deaths from cannibalism, starvation, disease, and territorial fighting, even though shooting, trapping, or poisoning is, of course, also inhumane, although less so than nature. To maintain healthy and properly stocked flocks and herds of livestock, they must be adequately managed and harvested. Eliminating the killing of livestock would result in their

Animal Rights Vs. Nature

vanishing from rangelands. The same is often true with deer and other wildlife where people have upset the original balance.

Regarding the imposition of suffering on sentient animals, H. Rolston III remarks that "...the strong ethical rule is this: Do not cause inordinate suffering, beyond those orders of nature from which the animals were taken." In modified environments I think we often have the moral responsibility of not even letting wild animals experience such excessive pain and suffering if we can greatly reduce it by regulating population densities through regulated hunting, trapping and other management tools. Interestingly, most people agree that when modifying an animal's environment by placing it in a zoo or home, we are morally obligated to ensure its welfare by providing food and other necessities.

Is it morally justifiable to hunt or kill wildlife when there is a surplus that the environment cannot support, or is it better to let them have what is frequently an agonizing "natural" death? Our self-righteousness about the sanctity of animal life often results in great cruelty. It seems to me that it is much better for game animals to be born and live a healthy life, under strict licensing regulations, if only to be later shot and trapped. The same applies to livestock and laboratory animals. In balanced ecosystems nature always sees to it that the annual surplus individuals are cropped or their habitats are damaged to the point that only a few animals can be supported until the habitat recovers. Why does the justifiable outrage that erupts when cattle or sheep are found

Hunting and Trapping

dead in a field due to inefficient management not occur when hundreds or thousands of surplus deer or other species die from inadequate harvesting?

Deer management in California is now governed too much by politicians and judges rather than by biologists supporting both the best interests of the game and of resource conservation. Consequently, each year a number of deer herds suffer tragically from starvation and disease, and twice as many of the state's deer are killed by cars than are shot by hunters. Female deer as well as bucks must be harvested as natural predators do in nature, where the ones that fall are mostly young. We must practice conservation and wise and effective use of natural resources, even though all such acts cannot be called humane. Sometimes this is the only way to sustain the biodiversity of many plant and animal species. The limiting negative impacts on wildlife today are not from hunting and trapping but from urbanization, roads, picnickers, hikers, campers, aquatic recreation, and other human activities that further limit suitable and essential habitat needs.

All animals are predisposed to overproduce, so their surplus must somehow be managed. When there is a surplus of herbivores, like deer, they tend to overbrowse and overgraze their habitat, with starvation, disease and parasitism then becoming important in reducing the size of the populations. Large carnivores are highly territorial so their surplus members must disperse or remain to be cannibalized or killed in combat.

Animal Rights Vs. Nature

The interest that ethical hunters have had of pursuing game has been the principal factor in nurturing conservation of wildlife and other resources. Now that hunting and trapping are regulated in North America, Europe, and other countries, it usually ensures sustainability and protection of the desired species and their habitats. Many species of birds and mammals have survived surprisingly well in North America primarily because they are game species. Current hunting and trapping of game species in developed countries for the most part are done ethically with regulations that prevent depletion and, as much as possible, inhumane treatment of the wildlife resource. There is still room for improvement among a minority of both hunters and trappers.

In contrast to species in nature, most people show great respect for the animals they kill. And these hunters have proven they will support conservation and self-regulation when these species are threatened with loss of habitat or other factors threatening their continued existence. Only people have evolved to the emotional state of having inhibitions and of feeling remorseful when they kill animals, even if for subsistence. It is surprising how some humanistic philosophers can condone coyotes disemboweling live sheep, and raptors and mammalian carnivores brutally killing prey, yet be opposed to regulated sport hunting. Remember, natural deaths in nature are often prolonged and very ugly. Skins and furs and meat are nature's natural renewable resource, so why not use them for clothing and food and reduce our consumption and the damaging effects by using so much of the finite supply of oil?

Hunting and Trapping

There are occasions where some wildlife professionals also use emotional propaganda with their congressmen. The profession was slow in pushing for adoption of nontoxic shot regulations for the hunting of waterfowl to eliminate the lead poison hazard to game birds and their predators, such as eagles. The wildlife profession should have responded sooner to this waterfowl mortality where lead shot is ingested or embedded in birds' flesh. When such delays occur, there is a danger that Congress will impose restrictions and that the professionals will lose control of the outcome.

A justified concern by some environmental groups is that wildlife managers tend to manage for "artificially" high populations of animals to benefit hunting, even though such management is also beneficial to many nongame species. Some of the less extreme groups have looked closely at how their former actions may ultimately affect wildlife populations and do seem to envision reduced population sizes in many cases, thus believing that such wildlife populations ultimately will be in better balance with the environment. I applaud the sincerity of these groups.

IS NATURE NEITHER RATIONAL NOR ETHICAL?

Since nature is so harsh and cruel, there is an important role for the hunter, trapper, and animal control agent to play in at least reducing this suffering by preserving a healthy balance amongst wildlife in human-altered environments.

Animal Rights Vs. Nature

Problem animals usually can be removed far more humanely by those means than would be their natural fate. If livestock operators managed domestic animals -- their sheep and cattle -- as badly as many state fish and game organizations are forced to under-harvest game animals such as deer herds, the livestock producers would not only be arrested but we would want them locked up for being so inhumane. In California, for example, we often have deer herds that greatly exceed the carrying capacity of their habitats. As a consequence, the survival of fawns is very low and the adults often succumb to starvation and diseases. Research has shown that nearly all does have twin fetuses following mating, but in areas of overpopulation the average fawn crop may equal less than one fawn per four does -- one eighth of what it should be. What has happened to the missing fawns? They either died before they were born or soon after birth due to malnutrition and/or diseases of their mothers, or from predators like coyotes and mountain lions.

Due to regulations, we now have little fear of hunter overkill and no worry about depletion of a species, but there is some concern about prolonged trophy hunting in game ranches causing a loss of desirable genetic traits. Hunters do their best to achieve clean kills, in contrast to natural predators. Clean kills are necessary to ensure recovery of the prey. Anyone with first-hand experience observing the feeding behavior of predators in nature will certainly have a good perception about the difference in killing by hunters and predators. Most hunting of game species is done ethically and in accordance with laws and regulations designed to prevent

Hunting and Trapping

depletion and inhumane treatment of the wildlife resource.

Except for professional wildlifers, few people recognize that it is the people who like to fish and hunt that originally saved the American wilderness. Conservation organizations are now helping. Wildlife populations, which man harvests by gun, trap, or rod, have literally never had it so good, because all animal populations must have high rates of predation of some type to remain healthy; otherwise, the surplus individuals are eliminated by inhumane and undesirable self-limiting (intraspecific) means, and the remaining animals experience excessive stress. In a balanced ecosystem, the surpluses of all species are cropped each year, one way or another, on a sustained basis. In modified environments, people can provide this needed predation, but it will be under strict licensing and regulations. States with wildlife management programs that include intensive hunting have dramatically improved the health of deer and other game populations. Even though wetlands are in short supply, in North America the conservation of wetlands would have fared much worse without the waterfowl hunters.

It may be comforting to someone to translocate or otherwise help individual animals, but such behavior usually is a selfish, not humane, act. Americans who insisted that India overprotect tigers and elephants are indirectly responsible for the more than one hundred lives of villagers lost annually to the uncontrollable surplus of animals now found in some Indian national parks. Similarly, when a rare or endangered grizzly bear, wolf, or lion leaves its sanctuary in the U.S., it

Animal Rights Vs. Nature

should no longer be considered an endangered species but should then come under that state's game and depredation laws.

Legal hunting and trapping do not cause extinctions; rather, more than any other source, these activities provide the funds to preserve habitats, hire wardens to protect the animals, and pay for biologists to ensure that healthy populations are sustained. Hunters find great satisfaction in the roles they play with nature; i.e., preserving the basic populations, managing a sustained form of production, and utilizing the natural recruitment of the resource. Respectful hunters do not kill for the joy of killing. It is interesting that only people have inhibitions and feel remorseful when killing; animals do not. Donations are needed to fund research to find a more humane alternative to steel-jawed leghold traps, although much progress was made in the 1980s.

Personally, I no longer hunt or trap, but my ethics clearly endorse such recreation-livelihoods for biological reasons. However, I believe we should be as kind as possible in the harvesting of game, hence should always search for the most humane way of carrying out these predatory roles. It will be necessary to inflict pain, but let's minimize the suffering. Once people modify an environment, they have a moral obligation to help regulate the balance of nature. Since we can respond to wildlife's needs in altered environments more rationally and ethically than can nature, we must be willing to serve as a predator. Animals are apparently born to be eaten by other animals in the natural scheme of things so

Hunting and Trapping

why not let the more humane sportsman help nature if done in an ethical spirit.

Those who claim killing of animals is a moral issue ignore the biological principles involved.

MOST OPPOSITION TO HUNTING IS PRIMARILY FOR PERSONAL REASONS RATHER THAN BASED ON BIOLOGICAL FACT.

How can one equate the suffering of animals that are shot, trapped, or poisoned with the suffering involved in slowly dying of starvation, disease, or any of the other vicissitudes of living that are often brought on between members of the same species? Isn't it possible that we can keep populations of wildlife, as with livestock, in a healthier condition through various harvesting methods than can nature, in what are no longer natural environments?

It seems to me that **SENSIBLE WILDLIFE CONSERVATION IS GRADUALLY BEING CROWDED OUT BY EXTREME ANIMAL RIGHTISTS AND HUMANE GROUPS TO THE DETRIMENT OF WILDLIFE.** There has been a very massive and effective attempt to propagandize the public into believing that wildlife resources are being driven to extinction by hunters, fur trappers, and fishermen. The media and the courts have paid too much attention to the half-truths, omissions, biased interpretation of data, and other

Animal Rights Vs. Nature

deceptive practices, probably because the real facts and truths were not being told by enough biologists. However, for the most part, the media are not interested in the wildlife managers' viewpoints because they have to be based on biological principles instead of propaganda.

Perhaps the greatest distortion of the balance of nature stems from anti-death groups who fail to recognize how easy it is to love a species population to death, or at least cause it to have a miserable existence as a result of overprotection by people.

MODERN SPORT HUNTING, FUR TRAPPING, AND FISHING HAS NEVER LED TO THE EXTINCTION OF A WILDLIFE SPECIES, AND WITHOUT SPORTSMEN MUCH WILDLIFE HABITAT WOULD HAVE BEEN LOST.

Game species have literally never had it so good, since sportsmen -- more than any other source -- provide money to protect the habitats and management of population densities of game species. It would make an interesting study to try and estimate just how many species of birds and mammals would be extinct by now if hunters and trappers had not, even if for selfish reasons, become organized to protect the animals and their habitats.

It is amazing how many people oppose the utilization

Hunting and Trapping

of animal skins, furs, and meat. No "game" animal in this country is endangered as a consequence of being so categorized. It is natural to recycle nature's wares. Skins and furs, which are not from endangered species, are renewable resources, and it is appropriate for people to use them. On the other hand, I don't think we have near the moral right to make our clothing from the finite supply of fossil fuel available, which we are using at an alarming rate, and at the expense of future generations. Hunters and trappers have not exploited the wildlife resources, but have, even if not consciously, often humanized the necessary harvesting of animal populations. Since natural predators enable their prey to exist at or near their carrying capacity, why shouldn't man help with hunting and trapping where natural predation has declined? Peregrines may be able to administer a 5-second <u>coup de grace</u> against prey birds, but wouldn't you agree that, for the most part, humans are usually a more humane predator of wildlife than most natural predators? Why does the public insist that livestock operators maintain healthy flocks and herds, yet object to proper harvesting of wildlife to keep populations healthy?

Hunting, trapping, and modern animal control practices can provide much better sustained stewardship and wiser, more enjoyable use of the wildlife heritage than the long-range effects of extremist groups. Any student of nature is surely aware of just how violent nature really is. Shouldn't we assist nature by acting as a predator where it is no longer possible for nature to reestablish the original predators in sufficient numbers to maintain a healthy balance and quality of life within the prey species?

The public needs an empathy for a realistic nature, not just for environments devoid of people. Special concern for individual animals is appropriate only if such action is not deleterious to a population of that species.

PEOPLE HAVE A MORAL OBLIGATION TO MANAGE NATURE ONCE THEY DISRUPT IT.

People must practice a realistic wildlife management and control ethic in **modified environments** rather than let nature be governed by her own uncontrollable brutality, which by now the reader will realize usually produces undesirable results. There is no place for a strong protectionist ethic in disturbed environments; it should be a management ethic. We must start teaching the real role of the balance of nature in disturbed environments and carefully explain why there is a great need for a death ethic where many individuals must somehow be removed to protect the welfare of the population. **SINCE NATURE DEMANDS A HIGH RATE OF PREMATURE DEATHS, WHY NOT HUMANIZE IT WITH MODERN HUNTING, FISHING, TRAPPING, AND ANIMAL CONTROL PRACTICES?** Life does not exist without death; there can be no life without death; death begets life.

In nature the single animal, which often needs to be

Hunting and Trapping

controlled, is biologically unimportant. The loss of a few individuals is insignificant. Nature is composed of populations. It is better to be inhumane to individuals if such action saves entire populations of that species. It is primarily through man's consumptive acts that the rate of premature deaths of wildlife can be reduced to preserve healthy reproducing populations and more desirable animal and plant communities. This is the objective of game departments. However, the regulation of population densities by hunting, trapping, and fishing requires discretion; it must be carried out in a biologically sound manner. **JUST BECAUSE SOME HUNTERS ARE NOT PROPER SPORTSMEN WITH RESPECT FOR ANIMAL LIFE, DOES NOT JUSTIFY CRITICIZING THEM ALL.** Similarly, because some animal welfare folks and animal rightists don't behave properly, doesn't mean that all supporters of these philosophies are bad people.

Wildlife conservation -- in fact, the wildlife profession -- is in trouble. There is a steadily growing percentage of the public that no longer understands the basic biological principles responsible for the so-called "balance of nature." The balance of nature issues, especially about wild birds and mammals, have become so emotional that even professional wildlifers find it difficult to discuss these matters objectively, and teachers, even scientific researchers, have become so emotional about birds and mammals that they usually can only say and write what they personally condone. Objectivity is gone; wildlife biologists too frequently preach the emotional subjects instead of stimulating others to think objectively. We are all caught in this emotional dogma.

Animal Rights Vs. Nature

In order to regain the public's confidence, it is essential that wildlife managers and biologists learn to speak more objectively, revealing all of the tradeoffs about the environmental issues in their field, including humaneness, whether or not they favor them. This is paramount if we hope to evolve healthy compromises instead of always losing to simplistic confrontation approaches. It is much better to negotiate and end up with a consensus than litigate. Fortunately, most of the public are not environmental extremists. We must be honest about the inevitable tradeoffs of any environmental issue if we hope to gain the confidence of the general public. **MY PLEA TO ALL BIOLOGISTS IS TO DISCUSS NATURE AND ANIMAL WELFARE ISSUES OBJECTIVELY, NOT EMOTIONALLY.**

To test yourself as to whether you are inclined to preach rather than teach objectively regarding the balance of nature, see how you would cope in front of a public group on many of the issues raised in this book. Even though living organisms are very valuable and provide a unique source of genetic material, could you leave ethical issues aside and objectively analyze whether or not it would really make such difference to the survival of the human race if a hundred, even a thousand, species of wild vertebrates were exterminated this year? I doubt it, because most people are emotionally conditioned to consider only arguments opposing the loss of species. It would be very difficult for most of us to enumerate reasons why the loss of one or many species might make little difference to human welfare or survival. Could you discuss how the eradication of some species might also help create

Hunting and Trapping

new gene pools for the remaining resident species? In other words, can you discuss <u>objectively</u> the impact of species removal on ecosystem dynamics, even though I hope that you, like me, really do not want to lose any fauna, flora, or ecosystems?

A. Miller succinctly put his finger on the basic reasons why wildlife biologists are usually not very objective in their analyses of emotional balance of nature issues that involve vertebrate wildlife species. Problem-solving behavior is influenced by deep-seated personality factors, including one's own emotional views. Therefore, environmental problem-solving cannot be viewed simply as an intellectual technical activity. It requires the education of the whole person. Miller found that the level of integrated thinking achieved by professional biologists may be limited by their moral values.

Too often we have distorted the reality of the balance of nature in fictitious, strident rhetoric, rather than objectively teaching the balance of nature. An examination of the scientific literature bears out the fact that, whenever a study about wildlife falls within this emotional arena, the author is careful to defend his/her emotional viewpoint, and any disapproved tradeoffs, no matter how real and obvious, are usually ignored. When people claim that man has destroyed any particular ecosystem, they rarely point out that such changes at the same time have created new habitats that favor other species.

Agriculturists have advanced much further concerning

managing healthy populations of ungulates than has the wildlife profession, which too frequently lets the density of wild animals exceed the carrying capacity in people-modified environments instead of managing both prey and predator populations where such is possible. Nature has seen to it that all organisms are obsessed with a breeding urge and are provided with the biological capacity to overproduce, thereby ensuring survival of the species. Since all animals are predisposed to overproduce, do you agree that a basic law of nature is that there <u>must</u> be a high premature death rate before almost any wildlife species can exist as a healthy population?

The human population of the world is overpopulated because science and technology have now made it possible for most babies to live long enough to reproduce. Only a few centuries ago this was not so. Death control for wildlife species also leads to overpopulation, disease, and starvation just as has happened with the human population in many parts of the world. For example, when an overpopulation of almost any species of herbivore is produced, either because man has stopped harvesting it or its natural predators have been removed or made inefficient by altering the habitat structure, it will overgraze and overbrowse its habitat and then decline in numbers due to disease, starvation, and other self-limiting factors. Any gardener knows it is necessary to thin radishes if plump, healthy radishes are wanted; similar types of predation of wildlife are equally beneficial to the well-being of their populations.

Hunting and Trapping

> TOO OFTEN PEOPLE INADVERTENTLY CREATE A "DEATH TRAP" FOR WILDLIFE BY ALLOWING THEM TO OVERPOPULATE.

Nature demands that many animals die prematurely to prevent tragic dieoffs of much greater number of individuals.

In recent decades, most human societies have developed a phobia against death and treat voluntary human deaths as obscene and illegal. This attitude must <u>not</u> be applied to wild animals as well. A healthy ethic, with deep ecological and moral conscience, would be to appreciate the glory of death in nature, for death means increased likelihood of a quality life for other individuals within that species. Biotic pyramids consist of food chains where all organisms feed on others and, in turn, are usually eaten. Recycling by nature necessitates that a surplus of animals be born and that only a few mature and reproduce, and we can humanize this recycling process.

Is it ethical not to offer humane assistance -- when it would be easy to do so -- just because the brutality or suffering occurring that you may want to prevent is natural? Where do we draw the line? For example, when weather conditions cause antelope and deer to starve, as happens periodically, should we feed them or let them die? When should we mercifully apply euthanasia to suffering animals? I believe man has an obligation to prevent natural brutality when he can, especially when induced by his actions, but only

Animal Rights Vs. Nature

when the long-range consequences to the ecosystem are considered. Feeding starving deer, or even large possible masses of starving people, frequently creates worse problems later on.

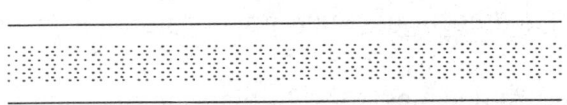

The following questions are to aid in testing your objectivity concerning the biological role of hunters and trappers.

1. As long as a species is not endangered, might it be better for people to follow nature's way and wear and use animal furs and skins instead of exhausting the finite supply of oil to make our clothes?

2. Since natural predators enable their prey to exist at or near their carrying capacity, why shouldn't man become a predator where natural predation has declined?

3. Do you think people can keep populations of wildlife in disturbed environments in a healthier condition through various harvesting methods than can nature?

4. Can you condone coyotes disemboweling sheep and the way raptors and carnivores brutally kill much of their

Hunting and Trapping

prey, yet still be opposed to regulated hunting and trapping?

5. How do you equate the suffering of animals that are shot, trapped, or poisoned with the suffering involved in dying of starvation, disease, or other vicissitudes of living brought on by competition between members of the same species?

6. Peregrines may be able to administer a 5-second <u>coup de grace</u> against prey birds, but for the most part aren't humans as predators more humane than most natural predatory species?

7. Are the reasons you oppose hunting based on personal or religious reasons rather than biological grounds?

8. Why does the public insist that livestock operators maintain healthy flocks and herds, yet object to a similar management of wildlife populations?

9. Do you agree that for the most part hunters and trappers harvest surplus animals that will die anyway, even when they seek older animals that make better trophies?

10. With reference to animal rights, do you believe that we should encourage wolves to reoccupy all of their former range throughout the United States and reintroduce grizzly bears into the Central Valley of California? Who is to decide how much of their former range such

Animal Rights Vs. Nature

species should be permitted to occupy?

11. Do you think it possible for large communities of people to coexist with migrating herds of buffalo, grizzly bears, wolves, or mountain lions?

12. Before poaching of elephants for ivory became so serious, the overabundance of elephants in some African national parks resulted in a great overutilization and destruction of the trees that elephants need as food. Do you think such lack of herd management was wise, or should such "surplus" elephants have been shot and utilized by the native people?

13. When species' populations are declining in an environment people have altered, is it better to try to manage the species or let nature take its course, when it will clearly be at the expense of the population's quality of life?

14. What liberties will you grant to the management and control of competing or predatory species if such action is needed to save an endangered species? To share game with hunters?

Even though the population of chital or spotted deer (<u>Axis</u> <u>axis</u>) has increased to where they overgraze and overbrowse the vegetation in Kanha National Park, India, the availability of this and other prey does not prevent dominant tigers from forcing surplus tigers to disperse out of the park and conflict with people. The same phenomenon will happen with grizzly bears, mountain lions, and wolves in the United States unless we prevent the development of surplus populations in their sanctuaries. (W.E. Howard)

It is not easy to design bear-proof containers that prevent bears from feeding on our garbage. But every solution has a trade-off, as the concrete foundation shown also provides safe harborage for ground squirrels, hence increasing the potential for rodent-borne plague. Six-inch wide wire mesh buried around edges of concrete prevents rodents from digging homesites beneath the concrete. (W.E. Howard)

8. FUR

I respect people's concerns about wearing fur coats, but I think they have been misinformed. Too many people start with the premise that NO furbearer should be taken regardless of how it is harvested. They are thinking emotionally and not from compassion for the penned or wild furbearers. Of course, many fur coats are worn only for vanity reasons. However, fur fashion is the only insurance we have that many furbearers will be born to live a quality life and eventually die humanely. On fur farms the animals are born only because their fur is wanted, not to provide food for other wild animals. About 60 to 80% of fur garments come from furbearers raised for this purpose, but furbearers are also wild-trapped. Many wild furbearer populations, such as muskrats and mink, also would not fare nearly as well if they were not being harvested. If their fur had not been wanted, many species would probably be extinct. Let me ask you, would you consider wearing furs if you were sure that such harvesting of the species actually improved the quality of the lives of these species and usually provided them a more humane death than wild furbearers experience naturally, or are you too emotionally involved?

The animal rights movement has had a positive value

Animal Rights Vs. Nature

of making people more conscious of the welfare of animals, but also has been quite successful in their potential goal of making hunting and fur seem vulgar and symbols of someone who is tasteless and uneducated. In addition to trying to make wearing fur unfashionable, they are striving for legislative mandates against trapping, whereas they really object to any use of fur no matter how humanely the animals are handled. They have been clever with emotional propaganda. In 1989 at least 40 bills were introduced in the U.S. to ban the use of leghold traps. They also have been skillful in obtaining "conscience" donations for their cause. What many people don't recognize is that most animal rights leaders also oppose the eating of meat or utilizing any animal parts NO MATTER how humanely the animals are raised and slaughtered, whereas in reality many furbearers live a good life just to become a fur coat.

IF YOU OPPOSE ANY EXPLOITATION OF ANIMALS, HOW DO YOU PROPOSE THE SURPLUS SHOULD DIE?

The counter argument against not wearing fur is that furs are natural products, and such recycling is what nature is all about. It is really wonderful for people that mink, foxes, and other furbearers produce such a valuable and useful product. We can't shear them like we do sheep, but the sheep must also eventually be killed. Some substitutes for real fur have the undesirable tradeoff of exploiting the earth's finite

Fur

supply of petroleum. The more desirable that fur becomes in fashion, the greater will the number of furbearers be that will be born to live a quality life that is longer than the average life of their wild furbearer counterparts.

Canadians of the Fur Institute of Canada, e.g., have made much progress in the development of more humane quick-kill trapping systems for furbearers. There are many other more humane traps. Those opposed to wearing furs object to any trap. However, they use the crueler out-of-date models of leghold traps in their propaganda to gain sympathy. Until effective alternatives are found, some leghold traps are still needed for many public problems of disease, danger, and pests. Nature demands that all furbearers eventually die, most prematurely, so isn't it wonderful that people can assist nature in this chore and be much more humane than occurs naturally?

NATURE KNOWS BEST; SHE WEARS FUR.

With game animals, pets, livestock, and captive furbearers, would it be better that they not be born to live a healthy safe life, just because they must die prematurely but humanely? If nature's ways are so perfect, then I suppose we should not protect game animals from life-threatening starvation, diseases, and other forms of suffering even if these traumas have been brought on because people are present.

Once a species becomes a legal game animal, the

populations of this species usually live a healthier life (before being harvested), because of licensing regulations paid for by hunters. Most of the mammals people see every day would not have been born if they had not been wanted, and many game species would be extinct or nearly so if sportsmen's organizations had not saved them. Wildlife are impacted far more by human activities other than trapping and hunting. Trapping is more humane than harvesting populations with cannibalism, starvation, or disease. We seldom realize that it is actually uncommon for an animal taken by a trapper to die with as much suffering as it would have experienced in a natural death, but more humane traps are needed.

    Most of the examples of trapping portrayed by those who are "anti" do not show the newer legal and more humane types of traps. However, even more humane furbearer trapping methods are still needed, and I wish animal rights and animal welfare organizations would do more to contribute to the needed research in this area instead of just raising money by being against trapping. No one likes the leghold steel trap, and vast sums of money have been raised to ban them. Yet I am not aware of any of the anti-trap organizations offering any financial support to aid research for finding a more humane way of harvesting furbearers. The leaders don't want any animals to be used by people no matter how humane the exploitation. It would be a demonstration of genuine concern about animal welfare if only a fraction of the money raised and spent trying to ban all traps was diverted to finding better ways of capturing these animals. I agree that the time has come where trappers should be certified so that they comply

Fur

with all necessary humane standards.

Neither wearing fur nor eating meat per se harms Mother Earth. With both, animal welfarism should be the main concern, not animal rights. If you think it is great for animals to be born, **EAT MEAT AND WEAR FURS.** Also, contribute to research for finding even more humane methods of harvesting wild furbearers.

In California it is the California woodpecker (<u>Melanerpes</u> <u>formicivorus</u>) that frequently pecks holes in shake roofs or the sides of wooden buildings to store acorns. The red-shafted flickers (<u>Colaptes</u> <u>cafer</u>) frequently become most annoying to home owners when they make nesting holes in the sides of their houses under roof eaves.

9. DOMESTICATED ANIMALS

Do you think it is wrong for lambs, calves, piglets, and chicks -- for that matter, also horses, dogs and cats -- to be born if they are never going to be allowed to run free? Remember, domesticated species are genetically programmed to depend upon humans for their existence and have no idea how their wild counterparts live. Animals that associate with man are freed of many of the constraints that otherwise keep their numbers in check. They are largely protected from hunger, unfavorable exposure to weather, disease, enemies, and intraspecific strife. How does one define the quality of life of domesticated species and animals "born" in a zoo? It is not easy because those living in the wild face many hazards.

Ranchers know that if they take good care of their livestock, the animals will take care of them. A livestock producer is much like a natural predator, only better because he or she can regulate both the age and the sex ratio of the "prey," and they must operate under many regulations. Also, they have compassion toward their prey, which does not exist in nature.

Animal Rights Vs. Nature

My personal view about the available space and treatment of farm animals is that in general farmers are strongly concerned about the health and well-being of their animals, which is contrary to what many urbanites think. Unfortunately, it is very difficult to determine the degree of suffering experienced by confined animals. Also, it is not simple to free chickens, cattle and pigs from the economics of intensive farming. To remain solvent, farmers need economic help to abandon intensive farming, brought on by frugal shoppers and the desire of agribusiness for profits. With a depressed agricultural industry, farmers cannot readily change these practices.

THE BUDGET SHOPPER HAS CREATED FACTORY FARMING.

The basic problem is economics, for there is almost no way that a farmer can adopt some proposed new animal management methods since to do so would make him unable to compete at the marketplace. If certain changes in farming practices were temporarily subsidized, so farmers could then permit farm animals to be raised more like they live in their natural environment, I suspect that such changes would come about more rapidly. In 1988 Sweden passed a stringent animal welfare program for chickens, cattle and pigs, but only time will reveal how successful or really necessary some of the restrictions were, for "factory farming" is not necessarily bad.

Domesticated Animals

Agriculture's problem of achieving long-term sustainability with reduced input is an unprecedented challenge. In many parts of the world there is also a need for agroforestry, the growing of trees and shrubs together with cash or food crops in a sustainable manner. To feed the world population, agriculture is trapped and must rely on inorganic fertilizers, agricultural chemicals and depletable energy resources such as fossil fuel. It is going to be difficult to convert to "sustainable" agriculture without loss of necessary productivity, although we must. Knowing that changes are inevitable, what kind of a planet do we want and how willing are we to make sacrifices to achieve our goals of "sustainable development"?

Is it better to have lived even if destined to die prematurely but humanely? Research is trying to measure how much domesticated animals may miss being able to roam free, scratch in dust, and various other behaviors they have never experienced. Since injured animals taken in by a wildlife care center are used to being free, they certainly must miss being free when held captive much more than caged chickens or veal calves. I do think we have a moral obligation to protect from nature's predation those animals we have domesticated, but how far does such "protection" extend? For example, dogs, cats, and livestock usually cannot peacefully coexist with coyotes or mountain lions in the United States (where even children are attacked by these carnivores). Also, domestic pets, e.g., house cats and dogs hunting for fun, cause significant losses to song birds and small rodent populations.

Is it inhumane for the modern western world to do

nothing to "better" the lives of the few remaining primitive tribes of people? Is it any more wrong for animals to live a healthy protected captive life than it is for humans to abandon living like primitive savages? What is quality-of-life for wild and domesticated animals? Should we provide primitive people with clean water, public health, education, and other western amenities? The primitive people would willingly accept many of our amenities. A caged chicken will scratch in the dirt for food or dust to control parasites. Is it cruel if they can't do it? When typical behavior performed by a species in the natural environment cannot be exemplified in captivity, does this deprivation cause some degree of suffering? If so, should all domesticated strains of animals be exposed to the same degree of predation, weather extremes, diseases, and hunger that their wild counterparts experience to keep them from "suffering"? Are we depriving ourselves by not being hunters like our ancestors, or were all humans that hunted in the past immoral brutes?

NEITHER PEOPLE NOR ANIMALS CONSCIOUSLY MISS ANY CONDITIONS THEY HAVE NO KNOWLEDGE ABOUT.

Ranchers, in particular, are acquainted with the gruesome sight of a 2-day-old calf with its hindquarters eaten away, yet still alive, or a sheep lying motionless and silent while a coyote consumes its intestines, to avoid being repeatedly attacked in the throat by the coyote. Coyotes and other

Domesticated Animals

predators of domestic animals must have their populations managed by hunters or other means for humane reasons, to protect the domestic animals that no longer possess their ancestral abilities to escape such predation. Since we all agree that native predators must be protected to maintain balanced ecosystems, it should be obvious that when we modify an environment and remove the top carnivores for human safety, people must serve as a predator to help nature continue its balanced system.

Sustained livestock husbandry, free of pest control chemicals or other pest management tools, is not possible unless the habitat can be modified so pest species cannot live there. Since domestic animals no longer possess many of their natural innate traits for escaping predators, perhaps they need an animal legal defense fund to give them better protection from coyotes, lions, bears and wolves.

"Bambi-ism" and self-righteousness about the sanctity of animal life can be responsible for great cruelty by ignoring the laws of nature. One would think, according to the philosophy of animal rightists, that we must be wrong when we protect our pets, livestock, and research animals from the types of life-threatening suffering their wild counterparts experience in the natural world. A philosophy which emphasizes protection of wild animals will often be contradictory to the necessities for quality of life of such animals.

Living in reference to all life necessitates a clear understanding of nature's survival-of-the-fittest death ethic, which

Animal Rights Vs. Nature

is what nature and her food webs are all about. Yes, nature can best be described as beautiful but brutal, because euthanasia and humane slaughter are not available in nature. If nature's ways are so wonderful, are we wrong in not letting game populations, pets, and domestic animals experience life-threatening starvation, diseases, climatic extremes, intraspecific fighting, exposure to predators, and other stresses common to their wild counterparts? All animals, including people must utilize other organisms for their own survival. The difference between humans and non-sentient animals is that we should be as humane as is possible in our use of animals.

It is not easy to develop strict guidelines that are consistent with stringent moral and ethical principles. The Animal Protection Institute of America suggests that people should eliminate meat and dairy products from their diets. This means no more beef cattle or dairy cattle could be born and live a good life compared to that endured by wild ungulates. I think this would be sad indeed, for I like all animals.

Raccoons (<u>Procyon</u> <u>lotor</u>) adapt well to living in residential areas with creeks, woodlots, fish ponds, and garbage and can become a nuisance when they eat valuable goldfish, tear holes in roofs looking for nesting areas, feed in the garden, or spill garbage. On farms they damage sweet corn and watermelons, young poultry and eggs, and other truck crops.

The domestic pigeon (rock dove) (<u>Columba</u> <u>livia</u>) is a species that even when feral lives closely with people and buildings, so much so they frequently become a serious nuisance and health hazard as well as cause economic damage to small grains and vegetables. They are only momentarily deterred from ledges by artificial snakes and owls, and only physical barriers and sticky chemicals will repel them from buildings.

10. ANIMALS IN RESEARCH

The morality of experimentation upon animals is not easy to appraise. Bioethical views are changing. One is no longer able to justify research with animals just to advance knowledge for knowledge's sake. But some misrepresentations must be corrected. Laboratory animals are rarely mistreated, neglected, or used frivolously by researchers. Researchers are usually compassionate with their animals and have invested many hours and resources in their study. Ethical scientists agree that animals have a right to be humanely treated, and they use alternative models to animals when available.

RESEARCHERS KNOW THEY NEED TO PROVIDE FOR THEIR ANIMALS' PSYCHOLOGICAL WELL-BEING

Researchers need healthy animals, hence provide for an animal's physiological and psychological well-being. They take all reasonable care to avoid undue suffering to ensure humane and responsible science. When possible, test animals are anaesthetized and then euthanized when the tests are terminated. Cruel treatment of animals in laboratories is rare and is not even remotely akin to what occurs in nature.

What is "pain"? Often it is the consequence of a volunteered action e.g., boxers, football players, cock fighting, dog fighting, and sexual combat of many species willing to invite conflict that is certain to be painful. Animals in the wild do not have morphine, euthanasia, or humane slaughter as do laboratory and farm animals. Family pets generally experience more human-inflicted suffering than do most research animals, and many pets are outright abused to the extreme in homes. The only way to abolish pain and suffering to all animals is to eliminate (exterminate) these species.

Many objections to animal research are rightfully targeted at cosmetic, drug, pesticide and other product research. I do not foresee the day when no animal will be needed in product testing, but advances in cell cultures, chemical tests and computer modeling is reducing the need for animal testing. And another breakthrough is the use of bacteria to identify cancer-causing agents. Similarly, in vitro tests are becoming an alternative to some traditional animal testing. There has been much recent progress in these issues.

To consider the ethics of using animals in research is relevant. I do not agree that animal experimentation is intellectual lethargy, scientific greed, or that the ethical mandates are perverted, but will agree that it is good that

Animals in Research

research design now comes under closer scrutiny. A utilitarian principle is that the results of research with animals is long-lived whereas the cost to the test animals is usually brief. Since we live in an economic world, it is obvious that research on animals may be done for economic benefit to the developers if it is popular and needed in the market place. However, much animal research is done by universities and government to benefit society. Such researchers are usually very conscientious.

Not only do wildlife serve as sentinals of environmental quality, but they share many infectious diseases with domestic animals and man. Do we have an obligation to continue conducting research involving animals just because such research has the potential for eliminating human and animal diseases and suffering? To those who need life-saving treatment, their moral decision is clear and unequivocal. Objecting to research with animals, yet wanting better health and a longer life, is contradictory.

I doubt if extreme animal rightists, should they have an accident or illness, will request that all medical treatments developed or tested on animals be withheld, because this means they would be denied anesthesia, pain killers, blood transfusions, insulin, antibiotics, vaccines, chemotherapy, CPR, coronary bypass surgery, orthopedic surgery, and reconstructive surgery. There are many more that we know animal rightists have used when they needed them.

Animal Rights Vs. Nature

LIFE-SAVING TREATMENT OF DISEASES
AND DEVELOPMENT OF DRUGS AND VACCINES
REQUIRE ANIMAL TESTING.

Animal testing has provided life-saving treatment for heart disease, AIDS, malaria, cancer, mental diseases, and other illnesses. Practically all drugs require animal testing. Vaccines for smallpox, rabies, polio, distemper, etc., all required animal testing. Research by behaviorists aids in the care of pets, farm animals, and the occupants of zoos. Most captive animals live much longer than their wild relatives. To prohibit the use of unclaimed (pound or animal shelter) cats and dogs for research and teaching only shortens the life of these "unwanted" animals. Furthermore, it necessitates breeding of more of these animals, even when a more uniform experimental breed may not be required. To suggest that most biomedical research does not require any experimentation with live animals is naive but let's reduce the animals needed to the minimum that will still protect people's health. There is a growing acceptance among researchers that alternatives to "painful" experiments on animals need to be one of the newer objectives of the scientific community.

WHEN POSSIBLE, ANIMALS ARE ANAESTHETIZED,
TESTED, THEN EUTHANIZED.

Animals in Research

It certainly is not cruel to anaesthetize an animal to put it to sleep for biomedical research or for the training of medical, veterinary or other students. Again, most of these laboratory animals would not be born and have a quality life if not wanted by science.

With man-animal relationships, there are no absolutist positions concerning man's use of animals, and it is only through research with animals that our humanity towards them has evolved. Even human rights to life and liberty are not absolute. It may be overridden by higher classes of society, capital punishment, killings for self-defense, or killing for national defense. It seems reasonable to require that humans treat animals as humanely as is feasible. Just because we may claim to use animals for human benefit or as a moral duty to advance human welfare, does not excuse any infliction of unnecessary pain and suffering.

OBJECTING TO RESEARCH BUT WANTING GOOD HEALTH IS CONTRADICTORY.

Stopping or seriously impeding use of animals in human and veterinary research laboratories will have a major impact on solving needed cures for people and animals. Animal rightists adamantly oppose the use of animals in research but gladly accept the benefits of such research and do not volunteer as research substitutes for animals.

Animal Rights Vs. Nature

Since so many in the animal activist anti-science movement are responsible for violent and illegal acts that have endangered both human and animal lives and destroyed property, they are being viewed as a threat to your and my freedom of choice. We want to be able to expect more research for the disease or handicap we may have or freedom as to what we eat or wear. It is easy to justify the social benefits of animal experimentation when it is for a problem you suffer, and easy to criticize if you are not affected. High-profile media and movie personalities have brought just one side of the issue to the television cameras. However, what animal rightists consider morality is now often at last being countered by the powerful political strength of industry.

Not many of today's scientists hold animals in low esteem. They are concerned about the welfare of their subjects. That such animals are just "research tools" has long passed. However, we are justified in stressing that researchers who experiment with animals must incorporate positive goals of animal welfare in their research protocol.

What right do animal rightists have to make the moral decision that people should not have pets and other animals? Ways are needed to achieve more objective dialogue between the conservative animal rightists and biomedical researchers. I do not think the animal rights philosophies have attempted to answer whether or not it is proper for genetic strains of laboratory mice and rats to be born, mate, have a high-quality life and then be humanly sacrificed. As a researcher, I eye with envy the millions of dollars raised each year by animal rights organizations. It would be a great help if only a small portion of this money could go toward research to develop alternatives to the use of animals in biomedical research and to find more humane ways of controlling vertebrate pests and harvesting wild furbearers.

ANIMAL RESEARCH HAS SAVED LIVES AND LESSENED SUFFERING OF BOTH PEOPLE AND ANIMALS.

It is surprising how many people don't understand, appreciate, or believe in the role of scientific research. A purported "guilty" researcher of liberated animals is a piece of cake for the media. Some claim that the greatest supporter of the animal rights activists is the headline-hungry media and those whose income is based upon contributions. The media arouse negative feelings, thus mold the perception of the public. Unfortunately, the media thrive on emotionalism and sensationalism and ignore biological truths. Too often bad news is "news" and good news unreportable.

Perhaps the most infamous and successful mammal is the rat (<u>Rattus</u> spp.). It has the genetic capability of adopting to a great variety of conditions. It does not need protection; we can't even get rid of it.

11. HUMAN POPULATION

The national population of China, or of India, today is greater and growing faster than the entire population of the world 200 years ago. Then the world's population was only 800 million, whereas in 1990 it is 5.3 billion, with 250,000 new people every day.

Practically all of today's world wildlife and environmental problems are indirectly the consequence of the rapid growth of the human population, hence the need for this chapter. During the first one or two million years of human existence, the human population averaged a growth rate of only one or two people a year, whereas the human population (5.3 billion in 1990) will soon be increasing by 100 million people a year (about 92 million in 1990).

No population can continue to increase indefinitely regardless of how much food there is. If civilization is to be viable, we must end the arrogant assumption that there are unlimited resources and infinite clean air and water and that we can survive our increasing pollution of the biosphere. The current world population of people has not only destroyed many of this earth's renewable resources but has seriously depleted such nonrenewable resources as soil, the supply of ground water, and the diversity of our fauna and flora. We used to think our ground water supplies were invulnerable to pollution, but they are now becoming contaminated, mainly by industrial wastes, which is making clean water a scarce and expensive commodity even for wildlife. Human population pressures force us to live in a "chemical" world, but somehow we must introduce greater precautions to protect wildlife as well as ourselves.

> HUMANS, LIKE ALL ANIMALS, ARE
> PRIMARILY CONCERNED WITH THE
> MOST IMPORTANT ANIMAL, THEMSELVES.

People must develop much greater voluntary restraint in reproduction, or conception will have to come under government control. Although such restraint tends to occur with advances in standards of living, there is a definite trend of a global decrease in living standard.

People are part of nature and must be accommodated. To survive, they must exploit the environment, as must all organisms. Skins, furs, and all wildlife are nature's natural renewable resource. Fortunately, under proper management we cannot only assist nature in utilizing there resources, but can do it more humanely than can nature. But man's appetite for resources and material things is devouring the earth's ground water and leaving polluted air, water and soil in a plundered planet. The environmental movement through the 1970s was largely directed toward keeping things pristine, whereas now much of the environmental focus is on public health issues such as chemicals and pests. The world cannot support an ever increasing human population armed with such potent destructive technology and human desires for even more materialism. We must become better caretakers of God's creation and improve our stewardship of the environment, or face an inevitable future catastrophic collapse. We desperately need a sensible environmental ethic in the national conscience that includes the laws of nature.

No population of wildlife or people can continue to grow beyond certain limits; eventually **involuntary** self-limitation in the form of premature deaths from starvation, pestilence (probably viruses with humans), and wars will prevent further increases in density. Since all finite space is limited, it is an indisputable fact that human birthrates and death rates must someday be balanced. Man must be aware of his dilemma, for if he attempts to feed the starving people regionally without effective control of the birthrate, he actually is only deferring to a later date the starvation of an even greater number of people living there. It is interesting to note that currently there is only a small surplus of grain in the world, and some populations can no longer produce enough food for themselves or afford to buy it from others. In contrast, part of the U.S. farm crisis of the 1980s was due to farmers having been led to believe there was a serious world food shortage, and hence an available market.

Animal Rights Vs. Nature

The world is facing this acute human overpopulation situation specifically because of advances in agriculture, public health, science and technology, and lack of similar progress and acceptance in the field of sensible birth control. Only centuries ago large families were necessary because of the high mortality rates. Families are not having more babies, but more now survive. While sex should remain an individual and private matter, procreation must become a public concern. For those who oppose abortion, they need to recognize that every surplus birth, even if it does not have a chance of surviving a normal life, utilizes resources before dying. Thus, with each premature death, the resources consumed by that child were still not available to others, hence the need for improved birth control. The same is true with wildlife. Also, not only has it become obscene and illegal for a human to die voluntarily, we are now interfering with the balance of nature by trying to prevent all wildlife from dying. Such animal right-to-life philosophy can seriously upset the balance of nature. Untold suffering amongst wild animal populations results when people try to prevent so many individual animals from dying naturally.

The obvious cultural goal for all societies wishing an abundant life and freedom from want should be a **LOW BIRTHRATE AND A LOW DEATH RATE**. All of the world's desperate needs -- ample food, permanent peace, good health, and a high-quality life -- are unattainable for all human beings both now and in the foreseeable future, not only because of economic, social, and political problems, but for one obvious reason: there are too many people. A soaring

Human Population

population means a shrinking of man's space on this earth. Hunger and overpopulation will not go away if we do not discuss them, and too many births are not just someone else's problem; it's everyone's concern. The destiny of overpopulation is erosion of civilized life as can be witnessed in many parts of the world, just as overprotection of wildlife can have devastating effects on the welfare of their populations.

The principal way in which man differs from animals is in his intellect, his ability to read and communicate, to learn, to use tools, his society, and his ability to regulate reproduction. He also differs from wild species in that he attempts to protect the unfit and all "surplus" births, which nature does not. Nature's evolution has seen to it that all organisms, including man, are obsessed with a strong breeding urge and are provided with the biological capacity to overproduce, thereby ensuring survival of the species. But before surplus animals and people die, they consume resources and contribute in general to other population stresses, all of which make the environment less suitable, thus lowering its carrying capacity for that species. Man needs space as much as do plants and animals. Wilderness areas and national parks will not be secure if the human population continues to grow. Even now to preserve natural areas, yet let people enjoy them, creates perplexing problems. Simple answers to environmental problems are hard to find because so many societal values tend to intrude.

Animal population densities are governed principally by the suitability of the habitat, interactions with predators, and species-specific self-limitation, but this is not so with people. The members of each wildlife species involuntarily prevent any further increase in their kind. This self-limitation consists of undesirable stresses that reduce the number of births or cause a compensating increase in death rates. Members of the population become their own worst enemies in the sense that they are then responsible for the increased rates of mortality and, perhaps, also some reduction in natality.

NATURE DOES NOT PRACTICE GOOD ANIMAL HUSBANDRY.

Nearly all organisms that are well adapted to their environment, except modern man, have built-in mechanisms for checking their growth before their necessary food supply and cover are permanently destroyed, or they are regulated by predation. But nature's population control processes are unemotional, impartial, and often ruthless -- a set of conditions that educated people will surely wish to avoid for humans even if we permit these processes to be inflicted on wildlife. Also, unfortunately, as with wild animals, when the human population level is below carrying capacity, the innate desire to have larger families then becomes very strong, making it difficult to observe wise "human husbandry" as done in livestock husbandry where population densities are carefully regulated.

Human Population

All nature's components are predisposed to overpopulate and, in fact, attempt to do so, thus causing a high rate of premature deaths, hence the high number of predators, scavengers, and decomposers found in nature.

As with wild animals, whenever man's population density has been markedly reduced through some catastrophe, or his technology has appreciably increased the carrying capacity of his habitat (environment), the growth rate of his population increases. With this accelerated growth, the population of people and wildlife then tends to overcompensate, temporarily growing beyond the upper limits of the carrying capacity of the environment. The excess growth of people (and wildlife) is eventually checked, however, by the interaction of a number of different kinds of self-limiting population stress factors.

Only self-limitation can stem the human population tide, and the only voice man has in the matter is whether it will be done **involuntarily** by nature's undesirable stresses -- as witnessed by the history of civilization -- or will be done **consciously** by having the population **decline** to an optimum carrying capacity. Man has transferred himself from being just a member of the ecosystem to a dominant position where he now, often mistakenly, assumes that the ecosystem is his to control at will. When this happens, he forgets that he is part of nature. People must see that their true place in the world is not to transcend nature but to discover and assimilate all they can about the truth of nature and their own role in nature. We must anticipate the soundness of the new balances we

Animal Rights Vs. Nature

create. It is, therefore, essential that we look back in history and extract those human behaviors which allowed our ancestors to survive millennia without long-term deterioration of ecosystems. Early man survived history by using the interest and not the capital of natural resources: hunting being the oldest profession. He could do this because his survival rates were so low that the total population of people was much less than now.

The daily economic pressures of individuals attempting to provide a decent civilization, especially for themselves, have led to the destruction of too many of the original ecosystems, sometimes ending up with desertification. As tragic as it may sound, when an underdeveloped country's population density is growing rapidly, both health and agricultural aid from the United States may not only be wasted but may severely aggravate an already deplorable social and economic situation in that country. As with wildlife, surplus populations of people cannot be indefinitely sustained. All species must either check their birthrates by various means or have a compensating mortality factor.

Insidious economic pressures seem to prevent any effective management of many resources in a manner that would provide for their utilization in perpetuity. Concrete and pavement surely are not the epitome of the human species' fulfillment even if they do prevent erosion of the soil underneath. An ecological appreciation of resource management is needed, and an ecological consciousness must replace the ecological atrocities. Further, we live in an economic world,

Human Population

and everyone -- including those who govern our wildlife resources -- is caught up in an economical, sociological, and political quagmire, due in large part to the explosive growth of the human population.

It is inevitable that the limited legacy of natural resources must steadily yield in the face of the current world population explosion. As the population swells, open spaces are inundated by a flood of housing, and resources shrink further. And, of course, wildlife suffers. The United States and other developed countries consume a disproportionately large share of the world's nonrenewable and other resources at an ever-accelerating rate, and finite resources are subject to eventual exhaustion. Fortunately, many environmental agencies have at least been successful in slowing down the process.

WHEN IT IS A MATTER OF HUMAN SURVIVAL, WILDLIFE PRESERVATION AND CONSERVATION HAVE LITTLE MEANING.

The human population first passed through the agricultural revolution, then the industrial revolution. These were followed by the public health revolution, and today the world is suffering from a revolution of rising expectations. Due to the selling pressures of business, and the exposure of goods and lifestyles by space satellites, people all over the world now have rising expectations. Nearly everyone wants more

Animal Rights Vs. Nature

material things and new experiences at the expense of resources. As individual aspirations rise and per capita resources fall, the widening gap between the " haves" and "have-nots" could well generate even more serious social and political pressures than already exist, with little thought of wildlife's welfare. The opening up of markets in Eastern Europe and Russia certainly will accelerate the continuous depletion of resources and will contribute considerably to existing pollution problems.

Technology and science can and do progress at an ever-increasing rate; but can social, political and religious views change rapidly enough to cope with this "progress"? Our intelligence is so great that it may destroy us because we seem to lack the wisdom and insight to recognize, and the ability to correct, what we are doing to ourselves, wildlife, the environment, and future generations.

It is truly a complacent society when it is unwilling to take any of the many steps available for preventing surplus births. Our primitive reproductive instincts cannot be condoned in the face of modern survival rates. The two are no longer in balance. To say that it is a basic human right to decide the number and spacing of one's children is to say that man may do whatever he wants to the environment without thought of its inevitable consequence to future generations. If other organisms do have rights, then such human behavior surely infringes on those rights, but in essence, we are undermining our own species.

Human Population

To achieve "quality living," with abundant wildlife, our ultimate goal must be a declining population growth rate <u>and</u> population size.

IF BABIES OF THE FUTURE ARE TO LIVE, THERE MUST BE FEWER OF THEM NOW.

The accelerated rate of soil erosion is one of the most serious crimes of civilization throughout the world, resulting in a steadily declining carrying capacity of the earth for both humans and wildlife. Yet it is interesting to note that in the United States we try to preserve all of nature's most eroded areas as national parks and monuments, e.g., the Grand Canyon National Park, with no attempt nor desire to dam their tributaries in order to stem any further erosion. We do this because we think we can still afford to lose all of this soil, and comfort ourselves by calling this "natural," the consequence of the balance of nature, in contrast to the deplorable soil erosion caused by agriculture, road construction, industrial and urban developments, and wild fires.

Other serious environmental problems threatening humans as well as wildlife are chemical pollutants that cannot be recycled, the pumping of ground water faster than it is replaced, the overutilization of natural resources, acid rain, increase of CO_2 and other chemicals in the atmosphere, and the list can go on and on. Nature can no longer "repair" on her own many of our misuses of natural renewable and nonrenewable resources because the pressures of the human population have become too great. Perhaps the most serious crime of herbicides, insecticides, fungicides, nematocides, rodenticides, avicides, and predacides is that they have enabled the human population in the world to increase so dramatically because these chemical tools have been so successful in protecting our health, plants, property, pets, and stored produce.

When surplus coyotes venture into suburbs seeking food (garbage, cats, small dogs, melons, etc.), some organizations want them live trapped (which is impossible) and then released in the wild (an inhumane act). It took two weeks of baiting with chicken to capture this hand-raised coyote, which had been put on a dry dog food ration. (W.E. Howard)

It often takes coyotes a number of minutes after crushing the trachea in the neck of a sheep before the sheep suffocates and can be thrown by the coyote. A typical coyote feeding behavior is then to start eating the small intestine while the sheep is still alive. The sheep seldom struggles for it would be subdued by another neck attack. Nature is not compassionate. (W.E. Howard)

12. CONCLUSION

Animal rightists are not asking for more humane treatment of animals, that is animal welfarism. The real goal of animal rights is to prohibit any exploitation of animals, especially the higher forms. The baseline with animal rightists is that even if some medical research with animals produced profound benefits to society, such research is still morally wrong. The animal rights movement is antivivisectionism in a different cloak. The animal rights movement has had the positive value of making people more conscious of the welfare of animals. But too often their activism has been extreme. This group is comparable to professional crusaders. They thrive on "anti" intellectualism and anti establishment. The animal rights movement is like the firebrand, radical, ecoterrorism form of environmentalism called "ecotage," for ecological sabotage. A compromise is required to reach middle ground between man's exploitation of natural resources and the total exclusion of human use of the resources.

No movement has a monopoly on virtue. With animal rights and some animal environmental movements, I bewail their extreme emotionalism and terrorism which do not help either animals nor the environment. The animal rights movement has entwined what is ethical with what can be used emotionally. They have tunnel vision with a bleak, empty vista without many animals surviving at the end. They resort to fabrication and fraud, replete with inaccuracies, misrepresentations, and outright false accusations. They are trying to be animal liberationists, at any cost. Some animal rightists would have you believe that all animals used in research are scalded, burned, strapped in place, injected, cut, dropped, wounded, bones broken, even bludgeoned, and all done in the conscious state. Animal rightists have been short on facts and long on slick media manipulation.

SOCIETY NEEDS TO UNDERSTAND NATURE'S DEATH ETHIC.

I do not ascribe legal rights to animals, yet I believe they have value, deserve respect and humane treatment; in fact, unnecessary pain and suffering should be prohibited. But who defines "unnecessary"? Deciding when pain and suffering are justified is more an issue of personal ethics than a set of moral principles or values, hence the conflict in viewpoints.

Many people readily believe that since animals were here first they have more rights to the environment than we do. We are considered intruders and not part of nature. Also, most of the unpaid volunteer zealots in this area are from cities where they are no longer exposed to nature in the raw

Conclusion

or witness the daily role of domestic animals. Fortunately, however, there are many good organizations promoting needed animal welfare and habitat preservation. The availability of suitable habitat is the key for wildlife. The problem that donors and I have in helping animals is to sort out which are the legitimate organizations. If you are a contributor to animal rights, check to make sure you are not inadvertently financing terrorism.

Animal rightists reject discrimination based on species and do not condone any treatment of animals that is not also appropriate to be done with humans. Yet no animal species recognizes that all forms of life are equal. Animal rightists fail to justify their actions, but many animal users also have not clarified their commitments to animal welfare.

A right-to-life philosophy that ignores the laws of nature can upset the balance of nature in man-modified environments with tragic consequences. It may be nature's way that animals naturally inflict pain upon each other, but it is incumbent of people to be as kind as possible to nonhuman life and avoid any unnecessary or deliberate mistreatment. Sometimes our sense of moral responsibility and conscience guides us against nature's way. But to go against nature's system just to accommodate our own emotions often means we will not be expressing true compassion for individuals or populations of a species.

Why can't people see the catastrophic problems in the environment resulting whenever a species loses its natural mortality factor and exceeds its carrying capacity? After all, look what too many people have done to the environment. Nature's death ethic is what the balance of nature is all about. Human social ethics need to be shaped around the laws of nature, not emotional propaganda, and our quest should be for environmental excellence.

Much of the sensational, heart-wrenching animal welfare propaganda we receive in the mail is a hoax, for it does little to help this or that animal, but merely improves the lifestyle of the solicitors, who really cannot afford to have these money-making issues resolved. Even though more humane techniques of harvesting wildlife are definitely needed, most animal welfare organizations never fund such research. They don't want a resolution of the problems. To them, even a compromise is unprofitable. Conscientious donors usually fail to notice the cheap-shots and half-truths that are being used, and how skillfully many of the basic laws of nature are circumvented. The more emotional is the solicitation for funds you receive in the mail, the less likely it is that your contribution will help animals.

TRANSLOCATING DISPLACED ANIMALS MEANS A CRUEL DEATH.

People's desire to save a displaced animal from death is usually prompted by one's conscience, not compassion for that animal. We want to give all animals another chance

Conclusion

whenever possible. When a surplus animal appears in your neighborhood or someplace else that is not a suitable habitat for it, one's desire is to capture the beast and release it where the species is known to be abundant. To do this naturally makes us feel good inside. Yet, when translocated in this manner, all released mammals experience untold suffering and cause disruption of the social structure of the populations where released. They seldom survive to find a vacant habitat and mate. Be humane; either euthanize them or eliminate them as humanely as possible with a kill-trap, poison, or gun. Translocating displaced mammals ignores quality of life.

All organisms must have a natural mortality factor or their numbers become regulated by self-limiting forces such as starvation, disease and cannibalism. Once people modify the environment and the top carnivores are eliminated for safety reasons, people must assist nature by being a predator. Nature demands meat eaters, and people predate under regulations so are more humane than natural predators.

Is it wrong for us to prevent our pets, farm and research animals from suffering like their ancestors did and do today in the natural world? Is it better for a laboratory or domestic animal to be born than not to have lived, when such animals usually suffer much less than do free-ranging wild animals? Trapping and hunting can have a positive impact on the health of an environment. What a sterile world it would be if we had no domestic animals, pets, and lost all the economic support of sportsmen to preserve game species and their habitats. The anti-hunting and trapping neurosis does not help animals. In

Animal Rights Vs. Nature

spite of the sanctity of individual animal life, the sacredness of the life of wildlife, the reverence for animal life, or societal concerns for animal life, it is still prudent to objectively analyze the ethics versus the potential long-range benefits of properly managing and controlling wildlife in today's modified ecosystems instead of overprotecting individual animals that then become overabundant and tend to destroy those parts of our wildlife heritage.

HEALTHY ANIMALS ARE PRODUCTIVE ANIMALS.

In modified environments the choice is ours: Let a survival-of-the-fittest new balance evolve, or help nature by maintaining the species and the habitat, even if it means being a predator to replace those displaced. Once people modify an environment, they have a moral obligation to help nature regulate the balance of nature. Since we can respond to wildlife's needs in altered environments more rationally and ethically than can nature we must be willing to serve as a predator. Wherever human conflict with wildlife exists and current control methods are not acceptable socially, the logical solution is to research for a more environmentally acceptable alternative means of helping people and animals coexist. Animal welfare concerns need to be incorporated in decisions regarding vertebrate pest management to avoid furthering the erosion of the public's confidence in professional animal control.

Conclusion

NATURE DEMANDS A HIGH RATE OF PREMATURE DEATHS; LET'S HUMANIZE IT WITH MODERN HUNTING, FISHING, TRAPPING AND ANIMAL CONTROL PRACTICES.

In nature the single animal that may need to be controlled is often of little biological significance as nature is composed of populations. The anti-hunters, inadvertently, seem to prefer that a larger number of animals suffer and die by nature's lingering, species self-limitation death, rather than have a smaller number die more humanely by the hunter and trapper. A strong case can be made that mankind could benefit wildlife populations far more by employing modern animal damage control principles than by being an animal preservationist or animal rightist adversary.

THE BELIEF THAT NATURE KNOWS BEST IN MAN-MODIFIED ENVIRONMENTS MUST BE DISPELLED.

Americans enjoy the cheapest and most beautiful food in the world. They are so well fed, they now have time to focus on food safety and humane treatment of farm animals. There is no question that farm animal welfarism is one of the most complicated issues agriculture faces. Unfortunately, the welfare of domestic livestock and the cost of production run counter to each other, but the demand for cheap food is still winning. There is no easy answer to minimizing production costs, yet at the same time maximize animal welfare. A reasonable compromise is necessary. Agriculture is responding, but it has lacked good bona fide evidence in support of traditional animal husbandry and slaughtering practices. In response to the animal rights movement, all agriculture organizations are now developing humane guidelines for their respective industries. This is good.

I don't expect people to instantly adopt my philosophy, but I hope this presentation will help others rethink their relationships with animals, since animals and the future sustainability of a viable and diverse natural resource base and public health are dependent on our capability to better manage these critical resources in an economical, social, and environmentally acceptable way for present and future generations. This cannot be achieved if the emotionalism of a few organizations dictates how we manage for both the human and wildlife long-term well-being. All of us need better communications on the animal rights issues. We need positive, constructive debates instead of name calling and increased emotionalism that further polarize the matter. Open forums and symposia are required. It is not a question of whether we should be more humane to individual animals in biomedical research, with domestic animals, pets, game animals, and furbearers, but rather how such a kinder treatment can be achieved.

In our pluralistic society, public debates on animal rights are urgently needed instead of further polarizing the

Conclusion

issues. Open forums that provide ample discussion time to permit the analyses of the broad spectrum of viewpoints might do much to clear the air on the animal rights movement and help us establish a clearer consensus on ethical and humane uses of animals. I approve of the deep, long-range ecological animal welfare movements over shallow environmentalism and animal rights activism. Perhaps the most successful counterattack against animal rightists is to label their animal rescue operations as "terrorist activities." This, of course, labels the perpetrators rather than focusing on the morality of the act. These acts are actually terrorism only when people are threatened, which has happened enough to make the label stick.

IT IS GREAT TO BE BORN, EVEN IF AS A LABORATORY, DOMESTIC OR GAME ANIMAL.

A 6-month-old black phase Australian brush-tailed possum (Trichosurus vulpecula). This species was ill-advisedly introduced into New Zealand as a fur resource. In national parks where possums are now found and with the help of introduced red deer (Cervus elaphus), a number of bush and tree species have been completely eliminated. It is also an expensive pest because it transmits bovine tuberculosis. (W.E. Howard)

13. REFERENCES

No attempt has been made to review the extensive literature on all the subjects this book covers. My personal library contains over 21,000 catalogued reprints. The following references were either cited or provide pros and cons about the animal rights issue.

Anon. 1990. Preserving the global commons. 13 articles. Phi Kappa Phi Journal, winter, LXX(1):2-45.

Brown, L.R., ed. 1990. State of the World. Worldwatch Institute, Washington, D.C.

Burger, W.E. 1983. Conflict resolution. Isn't there a better way? National Forum (Phi Kappa Phi Journal) 63(4):3-5.

Bywater, A.C., and R.L. Baldwin. 1980. 1. Alternative strategies in food-animal production. Pages 1-30 in Animals, Feed, Food and People: An Analysis of the Role of Animals in Food Production, R.L. Baldwin, ed. AAAS Selected Symposium 42.

Callicott, B. 1980. Animal liberation: A triangular affair. Environmental Ethics 2:320.

Carbyn, L.N. 1989. Coyote attacks on children in Western North America. Wildl. Soc. Bull. 17:444-446.

Case, R.M. 1974. Thoughts on the nature of naturalism. BioScience 24(5):307-309.

Cassell, E. 1989. What is suffering? Pages 13-16 in Science and Animals: Addressing Contemporary Issues, H.N. Guttman, J.A. Mench, and R.C. Simmonds, eds. Scientists Center for Anim. Welfare, Bethesda, Maryland.

Cohn, J.P. 1988. Captive breeding for conservation. BioScience 38(5):312-316.

Curtis, S.E. 1989. Animal welfarism and its impact on animal agriculture. The National Provisoner, April 22, 13-16.

Devall, B., and G. Sessions. 1985. Deep Ecology. Gibbs M. Smith, Inc., Layton, Utah.

Diamond, J.M. 1989. The present, past and future of human-caused extinction. Phil. Trans. R. Soc. Lond., B, 325:469-477.

References

Dubos, R. 1973. Humanizing the earth. The Rotarian 122(6):15-18.

Ehrenfeld, D.W. The Arrogance of Humanism. Oxford Univ. Press, New York.

Ehrlich, P.R., and A.H. Ehrlich. 1990. The population explosion. Amicus Journal 12(1):22-29.

Feirabend, J.S. 1984. The black duck: An international resource on trial in the U.S. Wildl. Soc. Bull. 12:128-134.

Fox, M.A. 1986. The Case for Animal Experimentation: An Evolutionary and Ethical Perspective. Univ. Calif. Press, Berkeley.

Garn, S.M., and W.R. Leonard. 1989. What did our ancestors eat? Nutr. Rev. 47(11):337-345.

Griffin, M.L. 1988. The legacy of Love Canal. Sierra 73(1):26, 27, 30.

Hennig, R. 1988. Was ist Jagd? [What is hunting?] Die Pirsch 5:3-6.

Hoagland, J.L. 1985. Infanticide in prairie dogs: Lactating females kill offspring of close kin. Science 230:1037-1040.

Animal Rights Vs. Nature

Howard, W.E. 1990. Nature and the rights of animals. Environs, Environmental Law and Policy Journal, Univ. Calif., Davis, 14(1):27-29.

Howard, W.E. 1990. Why lions need to be hunted. Proc. Mountain Lion Workshop, 3:66-68.

Howard, W.E. 1986. Nature and Animal Welfare: Both are Misunderstood. Exposition Press of Florida, Pompano Beach. (out of print).

Howard, W.E. 1984. Balance of nature: Fiction and reality. Pages 469-479 in 49th N.A. Wildl. and Natl. Res. Conf., March 23-28, 1984.

Howard, W.E. 1983. Livestock predators and the balance of nature. Pages 106-113 in 1983 Yearbook of Agriculture, J. Hayes, ed., USDA, U.S. Govt. Print. Office, 1983 0-416-273.

Hutchcroft, T. 1983. Responding to media cheap shots: Observations on the CAST experience. CAST Paper 16, Council for Agric. Sci. and Techn., Ames, Iowa.

Hutchins, M., V. Stevens, and N. Atkins. 1982. Introduced species and the issue of animal welfare. Int. J. Stud. Anim. Prob. 3(4):318-336.

Kellert, S.R. 1982. Striving for common

References

ground: humane and scientific considerations in contemporary wildlife management. Int. J. Stud. Anim. Prob. 3(2):137-140.

Leopold, A. 1948. A Sand County Almanac, and Sketches Here and There. Oxford Univ. Press, New York.

Lutts, R.H. 1990. The Nature Fakers. Fulcrum Publ., Golden, Colorado.

MacMahon, J.A. 1983. Nothing succeeds like succession: Ecology and the human lot. 67th Faculty Honor Lect., Utah State Univ. Press, Logan.

March, B.E. 1984. Bioethical problems: animal welfare, animal rights. BioScience 34(10):615-620.

McKibben, B. 1989. Reflections: The end of nature. The New Yorker, Sept. 11, 47-105.

Miller, A. 1982. Environmental problem-solving: Psychosocial factors. Environ. Manage. 6(6):535-541.

Midgley, M. 1985. Evolution as a Religion: Strange Hopes and Stranger Fears. Methuen, New York.

Midgley, M. 1984. Animals and Why They Matter.

Univ. Georgia Press, Athens.

Naess, A., and I. Mysterud. 1984. Philosophy of wolf policies. I: General principles and preliminary exploration of selected norms. Conservation Biol. 1(1):22-34.

Norton, B.G., ed. 1986. The Preservation of Species: The Value of Biological Diversity. Princeton Univ. Press, Princeton, New Jersey.

Novak, M.A., and S.J. Suomi. 1989. Psychological well-being of captive primates. Pages 5-12 in Science and Animals: Addressing Contemporary Issues, H.N. Guttman, J.A. Mench, and R.C. Simmonds, eds. Scientists Center for Anim. Welfare, Bethesda, Maryland.

O'Bryan, M.K., and D.R. McCullough. 1985. Survival of black-tailed deer following relocation in California. J. Wildl. Manage. 49(1):115-119.

Proulx, G., M.W. Barrett, and S.R. Cook. 1989. The C120 magnum: An effective quick-kill trap for marten. Wildl. Soc. Bull. 17(3):294-298.

Regan, T. 1986. Animal Sacrifices: Religious Perspectives on the Use of Animals in Science. Temple Univ. Press, Philadelphia, Pennsylvania.

References

Regan, T., and P. Singer. 1989. Animal Rights and Human Obligations. Prentice Hall, Englewood Cliffs, New Jersey.

Register, U.D., and L.M. Sonnenberg. 1973. The vegetarian diet. J. Am. Diet. Assoc. 62:253-261.

Richards, R.J. 1988. The Evolution of Morality. Univ. Chicago Press.

Rollin, B.E. 1981. Animal Rights and Human Morality. Prometheus Books, New York.

Rollin, B.E. 1989. The Unheeded Cry: Animal Consciousness, Animal Pain, and Scientific Change. Oxford Univ. Press, New York.

Rolston, H., III. 1986. Philosophy gone wild. Prometheus Books, New York.

Rolston, H., III. 1988. Environmental Ethics: Duties to and Values in the Natural World. Temple Univ. Press, Philadelphia, Pennsylvania.

Rowan, A.N. 1984. Of Mice, Models, and Man: A Critical Evaluation of Animal Research. State Univ. of New York.

Rowan, A.N., Ed. 1988. Animals and People Sharing the World. Tufts Univ., Univ. Press of New England,

Hanover, New Hampshire.

Singer, P. 1985. In Defense of Animals. Blackwell, New York.

Singer, P. 1975. Animal Liberation: a New Ethic for Our Treatment of Animals. New York Review, New York.

Soulé, M.E., ed. 1986. Conservation Biology: The Science of Scarcity and Diversity. Sinauer Assoc., Sunderland, Massachusetts.

Sperling, S. 1988. Animal Liberators: Research and Morality. Univ. Calif. Press, Berkeley.

Taylor, P.W. 1986. Respect for Nature: a Theory of Environmental Ethics. Princeton Univ. Press, Princeton, New Jersey.

Tans, P.P., I.Y. Fung, and T. Takahashi. 1990. Observational constraints on the global atmospheric CO_2 budget. Science 247:1431-1438.

Timm, R.M. 1982. Teaching vertebrate pest control: a challenge to wildlife professionals. Trans. N.A. Wild. and Natural Res. Conf. 47:194-199.

Uvarov, O. 1985. Research with animals: requirement, responsibility, welfare. Laboratory Ani-

References

mals 19(1):51-75.

Vaughn, C. 1988. Animal research: ten years under siege. BioScience 38(1):10-13.

Wilson, E.O., ed. 1988. National Forum on Biodiversity. Natl. Acad. Press.

Zaslowsky, D. 1988. Poverty, prosperity, and preservation. Sierra 73(1):24-26.

14. APPENDIX*

ANIMAL RIGHTS IN PERSPECTIVE WITH NATURE

Dr. Walter E. Howard

INTRODUCTION

Many people are torn first one way then the other by the conflicting arguments pertaining to animal use.

A positive value of the animal rights movement is the increased consciousness most people now have about the welfare of pets, livestock, and laboratory animals.

All of nature's animals, wild or domesticated, have legitimacy and value, and so do people.

---

*Distributed to California State Legislature on 21 April 1989 by Assemblyman Chris Chandler.

Animal Rights Vs. Nature

Animals are not human, but they do have "feelings."

We cannot ignore the profound question being raised about what, if any, is the difference between people and animals. To say that animals are not "human" does not resolve the issue.

It is natural for some people to enjoy animals as pets, beasts of burden, for sport, and for food and products; otherwise, most of these animals would not have been born.

I do not ascribe rights to animals, yet I believe they have a moral value, deserve humane treatment; in fact, unnecessary pain and suffering should be prohibited. But who defines "unnecessary."

What is right or wrong concerning the treatment of animals is often in the eye of the beholder and is really determined by one's personal ethics.

The balance of nature requires meat eaters.

All organisms must have a natural mortality factor or be regulated by species self-limiting factors such as starvation, disease and cannibalism.

NATURE

To really be helpful to wild or domesticated animals, one must

Appendix

understand and appreciate nature.

People are part of nature, and they are not going to go away.

It is an axiom that man must exploit the environment and deliberately unbalance it to survive.

Sometimes when we express our love and pity for animals or a local population of species, we are inadvertently very inhumane even though our conscience may feel good.

Nature controls population densities with predation, starvation, disease, cannibalism, territoriality, infanticide, etc.; euthanasia or humane slaughter is not available to them.

Animals can be cruel in their exploitation of others.

How can we say that animals are self-conscious, i.e., aware of themselves or conscious beings, and then condone their brutality to other species and often their own kind?

Is it wrong to prevent our pets, farm and research animals from suffering like their ancestors did or still do in the natural world?

Living in reverence of all life requires an understanding of nature's survival-of-the-fittest death ethic.

Nature can be tender and delicate, but is also a battlefield where often bizarre types of cruelty are inflicted on animals.

Animal Rights Vs. Nature

If animals act intentionally with a conscious will, we must accept that nature is cruel and is dependent upon some form of acute suffering of all animals.

Nature is often a tooth-and-claw blood bath where many animals eat others and in turn are eaten--the survival of the fittest.

No animal populations can be healthy without a high premature death rate.

Animals may be intelligent, sentient beings of inherent value, but only people, the dominant animal, has conscience and shows compassion for other species.

Nature is composed of populations of organisms, and single animals are rarely of biological importance; therefore we are justified in being inhumane to a few individuals to save a population of the species.

A right-to-life philosophy can upset the balance of nature with tragic consequences to overprotected populations of animals.

Nature is wonderful! Salmon spawn once, then literally rot to death if not killed.

All species are programmed to overproduce; the surplus must die prematurely.

Nature's food-web is based upon an overproduction of all

Appendix

organisms.

People must help nature maintain a healthy balance of organisms in altered environments.

In modified environments the choice is ours: let a new balance develop or help the species we consider desirable.

ANIMAL RESEARCH

Views are changing. It is no longer easy to justify research with animals just to satisfy a scientist's curiosity.

Is it better for a laboratory or domestic animal to be born than not to have lived, when such animals usually suffer much less than do free-ranging wild animals?

Is one justified in sacrificing a few animals to learn how to keep many others from much suffering and premature death?

Sometimes the public overlooks that knowledge and progress in animal research often is the slow process of piecing together the evidence laboriously collected by many people over long periods of time.

Mainly as the result of research with animals, people in the U.S.A. now live an average of 20.8 years longer (U.S. Department of Health and Human Services).

Animal Rights Vs. Nature

Animals are used in research and teaching because there are no effective alternatives to the living organism.

The practical benefits from much research with animals is never easy to predict or overtly justify.

Unfortunately it is difficult to obtain much scientific knowledge without using animals.

Animal research has saved lives and lessened suffering of both people and animals.

Animal testing has provided life-saving treatment for heart disease, AIDS, malaria, cancer, mental diseases, and other illnesses.

Practically all drugs require animal testing.

Vaccines for small pox, rabies, polio, distemper, etc., all required animal testing.

Ethical biomedical researchers use alternatives to animals whenever they are available.

Ethical scientists agree that animals have a right to be humanely treated.

In research (when possible), test animals are anesthetized, and euthanized when tests are terminated.

Appendix

Animal researchers are usually compassionate with their animals and most enjoy the mental and spiritual rewards of nature appreciation.

Responsible researchers know how important it is to provide for an animal's psychological well-being.

The ethological findings of research by behaviorists can aid in the care of farm animals, pets and zoo animals.

How can an animal <u>miss</u> a freedom it has never experienced?

Laboratory animals and livestock are usually NOT mistreated, neglected, or used frivolously by responsible researchers and farmers. Of course there are exceptions.

FARM ANIMALS

Do you think it is wrong for lambs, calves and chicks, for that matter also horses, dogs, and cats, to be born if they are never going to be allowed to run free?

Domesticated species are genetically programmed to depend upon humans for their existence, and I do not consider such actions by people as "moral blindness" when man is protecting them from the suffering and brutality of nature.

The "quality of life" of animals born in captivity that do not

Animal Rights Vs. Nature

know what they are missing is difficult to define.

Animals that associate with man are often freed of many of the constraints that formerly kept their population density in check. They are largely protected from hunger, unfavorable exposure to weather, disease, enemies, and intraspecific competition.

With some species, e.g., the horse, their domestication by man may have saved them, as horses had almost vanished by 2000 B.C.

Economical ways of "enriching" the environment for caged and penned animals are needed.

It is not easy to free chickens, cattle, and pigs from the economic restrictions of factory farming.

Farmers need economic help before they can abandon intensive factory farming brought on by frugal produce shoppers.

Is it better not to have lived if you are going to die prematurely but humanely? How about with livestock, chickens and captive furbearers?

Just because a chicken may look happier to us if it is able to scratch and dust in soil doesn't mean it is suffering if confined in a cage having never experienced running about on the ground. Data are needed.

Appendix

We buy pesticides in drug stores to control pests "in and on our bodies."

Do we have a moral obligation to protect from predation animals we have domesticated?

HUNTING AND TRAPPING

Self-righteousness about the sanctity of animal life can be responsible for great cruelty to animals: i.e., overprotection can result in dieoffs from disease and starvation.

Opposition to hunting and trapping is primarily for personal ethics rather than being based on biological facts.

In modified environments where the original predator-prey balance no longer exists, nature often needs people as "predators" to prevent excessive deaths from cannibalism, starvation, disease, and territorial fighting.

How can one insist that livestock operators maintain healthy and properly stocked flocks and herds, yet object to harvesting wildlife to keep populations healthy?

How can humanistic philosophers condone coyotes disemboweling live sheep and raptors and carnivores brutally killing prey yet be opposed to sport hunting?

Animal Rights Vs. Nature

Furbearers and game usually die with much less suffering when harvested by people.

If nature is not guilty for brutality, then death by human predators can be considered part of the wholeness of life.

It is more humane to regulate wildlife population densities by hunting and trapping than having a much larger number succumb to disease and starvation.

A philosophy that emphasizes a reverence for life of animals may be contradictory to the necessities for quality of life.

People often show respect for animals they must kill, but this is not the case when animals kill other animals.

Animals do not have inhibitions and feel remorseful concerning the killing of other animals.

Before modern game management laws and practices and the Endangered Species Act, some species were even exterminated.

Illegal harvesting of game is still a problem, but much less so because of state and federal wardens, largely paid for by hunters and trappers.

Game species have literally never had it so good now that sportsmen pay for their protection and habitats.

Appendix

Most state fish and game conservation agencies are probably too conservative with their harvesting regulations.

There has been a tendency for wildlife managers to manage some game for "artificially high" populations of animals to benefit hunting.

No wildlife resource is being driven to extinction by legal hunters or fur trappers as is often claimed.

Skins and furs are nature's natural renewable resource. Let's use them and not exhaust the finite supply of oil.

ANIMAL RIGHTISTS

Nature's demand for a high rate of premature deaths can be considerably humanized with modern hunting, trapping and animal-control practices.

In modified environments true harmony between people and the fauna can be established only through management of animal populations.

Wild animals seldom show compassion for other species as humans often do in their efforts to achieve humane killing.

Few recognize that virtually no agricultural crop could coexist in a productive condition with free-ranging native mammals.

Animal Rights Vs. Nature

Kelp is probably an exception.

Is rodent-proofing your house denying rodents their rights?

Are we supposed to tolerate rats, fleas, bedbugs, and ants in our houses?

How can we equate the suffering of animals that are shot, trapped, or poisoned with the suffering of slow death from starvation or disease?

In man-altered environments animal control agents can greatly assist the protection of endangered species and preserve a healthier balance among animal populations.

There is a great need for an alternate to the offset steel-jawed, leghold trap, which is still needed for wary coyotes.

Without effective pest prevention, pest control measures are inevitable.

Vegetarians eat almost exclusively unnatural exotic plants. They probably could not survive on just native plants.

To grow edible vegetables, fruits, or nuts, it is necessary to destroy natural communities that originally occupied those habitats.

Why should people become vegetarians? Nature demands meat consumers to preserve the balance of nature.

Appendix

Animal rightists oppose any exploitation of animals and try to protect the interests of animals that are able to feel pain and suffering.

Animal welfare is the expression of kindness and concern for the welfare of animals people use, whereas animal rightists oppose all use of animals and think the human race should become vegetarians.

Objecting to research with animals, yet wanting better health and a longer life, are contradictory.

Are our morals and ethics all unnatural, for they are merely rational views people have developed primarily in recent history?

Domestic animals are legally the chattel or property of the owner, but the animal rights movement is out to change this.

The emotional concerns of animal rightists do not necessarily mean their actions are humane.

If "higher animals" are self-conscious beings, they are certainly brutal, blood-thirsty and noncompassionate beasts.

When pain and suffering are minimized, do animal rightists still oppose use of animals?

Some feel that animals are inherently entitled to the same constitutionally afforded rights as we are.

Animal Rights Vs. Nature

The roots of the aversion to killing of animals are partly in the current urbanized "Bambi" society that has lost touch with natures's life-and-death roles.

Is it wrong to deny rats and ants entry to your home?

Should primitive people, with low infant survival and high premature death rates, be westernized to improve their quality of life? Are they like battery chickens and veal calves?

Injured animals taken in by a wildlife care center may suffer more than battery chickens or veal calves because they are used to being free to fly or run. An animal may not miss a freedom it never had.

Animal rightists are abolitionists opposing any "use" of animals, apparently because they do not understand nature, whereas animal welfare opposes unnecessary pain and suffering.

Animal rightists seem to consider themselves judge and jury and use their dogma to justify disobedience and terrorism.

If it is wrong to permit laboratory animals to be born, it must be sinful to let unwanted human embryos become children.

Since game cocks fight voluntarily, is it wrong for people to permit them to fight? Such antics are common in nature. However, in nature they do not have metal weapons attached to their spurs to inflict deeper wounds.

Appendix

Animal rightists should make sure the actions they take are truly compassionate. Domestic, laboratory, and zoo animals often live much longer than their wild relatives.

Animal rightists feel that prolonging the life of a laboratory animal is more valuable than obtaining knowledge.

Many "anti" activists seem to be professional crusaders who would be lost if their cause was resolved. Perhaps me also.

The animal rights movement has stimulated a great deal of soul searching, which is positive, but now they have become much too extreme in their activism.

Animal rightists are comparable to professional crusaders and thrive on "anti" intellectualism and establishment.

Will a dedicated animal rightist offer his or her body to science in place of a rat's?

Are rats equal to people?

The animal rightists I know kill ants that come into their kitchen.

What type or respect do animal rightists give dead animals?

What are animal rightists' views about the merciful killing of injured animals and the killing of terminally ill people? Does quality of life enter the decision about terminating an unwanted embryo?

CONCLUSION

The current struggle most of us are experiencing in rethinking our relationships with animals is good. I have tried to be objective, but as with all humans, I am inherently biased.

To effectively resolve the conflicts requires both sides to face and discuss the differences that divide them, no matter how uncomfortable it may be to acknowledge the concerns of the other.

Is it a waste of time to strive for consensus where extremism is prevalent?

We live in an economic world but still lack the needed economic incentives for conserving wildlife and other environmental resources.

As the human population grows, with the dependency people place on domestic plant and animal species, the number of wild species that can share the finite environment will decline.

We must learn to be fair, humane, and even loving to animals, yet use them for our benefit.

The enthusiasm of animal rights extremists needs redirecting

Appendix

from their clandestine and covert activities against others who may exploit animals.

Once man has modified an environment, he has a moral obligation to manage and control the wildlife species present.

In modified environments, people can respond to wildlife's needs more rationally and ethically than nature.

We have a moral obligation to manage nature as best we can once we have disrupted it.

By same author:

Nature and Animal Welfare: Both are Misunderstood. 1986. Out of print.

Nature's Role in Animal Welfare. The Hume Memorial Lecture, presented at The Royal Society of London, 29th November, and The Royal Society of Edinburgh, 1st December, 1989. Sponsored by Universities Federation for Animal Welfare (UFAW).